CLEAN CHEAP HEAT

CLEAN CHEAP HEAT

The Development of Residential Markets for Natural Gas in the United States

John H. Herbert

PRAEGER

New York
Westport, Connecticut
London

Library of Congress Cataloging-in-Publication Data

Herbert, John H.
    Clean cheap heat : the development of residential markets for
natural gas in the United States / John H. Herbert.
      p. cm.
    Includes bibliographical references and index.
    ISBN 0-275-94204-X (alk. paper)
    1. Gas, Natural—United States—Marketing—History. 2. Gas
industry—United States—History. 3. Dwellings—United States—
Heating and ventilation—History. I. Title.
HD9581.U5H47 1992
33.8'23312'0973—dc20   91-35784

British Library Cataloguing in Publication Data is available.

Library of Congress Catalog Card Number: 91-35784
ISBN: 0-275-94204-X

First published in 1992

Praeger Publishers, One Madison Avenue, New York, NY 10010
An imprint of Greenwood Publishing Group, Inc.

Printed in the United States of America

The paper used in this book complies with the
Permanent Paper Standard issued by the National
Information Standards Organization (Z39.48-1984).

10 9 8 7 6 5 4 3 2 1

CONTENTS

TABLES AND FIGURES

Tables

Figures

PREFACE

During the 1970's and 1980's I read many analyses of natural gas markets in which comments were made about the historical development of residential gas markets. In many instances, an examination of primary data and other evidence revealed that the author of the statements had been misled. Eventually, I realized that a comprehensive analysis of historical data on this market was needed.

In my examination of evidence during leaves of absence from work I was surprised to find how dramatically gas appliances had changed economic and social conditions of many households; how differently industry and government representatives had interacted in forming the market depending on the time period; how quickly the market developed in some places and periods and how slowly in others; and finally, how fragile available estimates of the effect of price on residential natural gas use were. These estimates were of particular interest because they had influenced investments of firms and government regulations and, thus, they had influenced the development of the market and the economic welfare of consumers.

The writing of this book was inspired by support from many quarters. Most of all I would like to thank my father and mother who gave me a deep appreciation for the importance of history, and my son, John, who seems to have a very natural love of history. But inspiration is one thing, work is another.

I would like to acknowledge and thank M. Elizabeth Sanders of the New School for Social Research; Harry Trebing of Michigan State University and Director of the Institute of Public Utilities; Mark Rose of Michigan Technological University; Noel Uri of the United States Department of Agriculture; Erik Kreil of the United States Department of Energy; and Phil Kott of the Bureau of the Census. I am grateful for their helpful comments on early drafts of chapters and for their encouragement to continue. I would also like to thank Lucia Di Venere, Jill Lady, and Ellen Herbert for editorial and other support. Nonetheless, the views and any remaining errors are solely those of the author.

CLEAN CHEAP HEAT

INTRODUCTION

During the 1950's households in Great Britain and Germany were almost totally dependent on coal distributors for satisfying their heating needs. Across the ocean in the United States, a relatively cheap, clean source of energy - natural gas (NG) - was being supplied to more households than any other fuel. For more than a century, residential natural gas (RNG) markets, formed by the interaction of regulatory bodies and large businesses, have significantly improved the physical, social, and economic well-being of households in the United States. Yet this book is the first comprehensive analysis of the significance of this market.

The development of RNG markets in the United States is one of the great economic growth stories of the twentieth century. The volume of sales in this market grew fifty-fold between 1906 and 1970 and eight-fold during the great growth period between 1945 and 1970.[1] Today, NG distribution companies are able to consistently satisfy the highly seasonal demand of households for energy in every part of the United States.[2] Things were not always this way.

HISTORICAL DEVELOPMENT

In the first decade of the twentieth century, most NG use in households occurred in western Pennsylvania, northern West Virginia, eastern Ohio, and southwestern New York. This region was the Middle East of its time, with abundant supplies of readily available oil and NG. By 1930 sizable markets had developed in the six south central states of Arkansas, Kansas, Missouri, Louisiana, Oklahoma, and Texas and in the four western mountain states of Wyoming, Utah, Montana, and Colorado. California had grown to be the fourth largest RNG consuming state by 1930.

Growth during the 1930's was much slower than during the 1920's due, no doubt, to the general decline of the economy. Nonetheless, the RNG market continued to grow during the Great Depression.

Despite the growth in the RNG market up to 1940, during the Second World War the major sources of energy for the household, especially for space heating, were still coal and wood. Wood was primarily used in the South to heat houses. Coal, with its heat value much greater than wood on a per-pound basis, was used to heat houses in the colder North. The reliance on wood and coal had many costs for households in addition to the dollar cost of the fuel. There were labor costs associated in some instances with gathering the fuel and in most instances with feeding furnaces, fireplaces, and stoves. There were also health costs associated with by-product particulates created both inside and outside the house. During the 1930's employment outside the household was frequently difficult to get, and periods of employment were often followed by spells of unemployment. Thus, household labor to support the use of coal and wood was available. But during the war years the number of workers employed by businesses increased as did the average number of hours worked per day. Less time was available to household members to perform chores around the house. Value of leisure time increased as money incomes increased, and the number of hours available for daily leisure declined. Hence, the value of lost opportunities for leisure or the opportunity cost of maintaining a wood or coal appliance in the household increased. This change in circumstances made labor-saving NG furnaces and water heaters increasingly attractive.[3]

Prior to the war years, the average cost of NG to residential customers changed dramatically at times, and there was a rising trend in NG prices in many cities and towns. A disparity in the price paid by residential customers relative to industrial customers frequently brought the charge of discriminatory pricing.[4] The average price of NG to industrial customers could be one-tenth the average price to residential customers. It was argued that a portion of this differential in price was a consequence of unregulated commerce in NG between states.[5] Such charges were used to gain support for the passage of the Natural Gas Act in 1938.

The baby boom after the war years also increased the opportunity cost of employing wood and coal in the household. More time and a cleaner household environment could be made available for the raising of children by the addition of a NG furnace or NG appliance into the household.[6] Moreover, the out-of-pocket cost of using NG to heat households was actually less than the cost of coal in such major cities as Detroit, Cleveland, and Cincinnati.[7] The price of NG in constant dollars also began to decline after the war years as improvements in pipeline and storage technologies reduced the overall cost of supplying NG. With improvements in pipeline and storage technologies, utilities in the northern United

States increasingly were purchasing relatively inexpensive NG from Texas, Louisiana, and Oklahoma throughout the year. NG was placed in storage during the summer trough in demand and then withdrawn from storage during the winter peak, when the demand for NG was frequently ten times greater than it was in the summer.

Although western Pennsylvania developed early as a major market, it was not until the 1950's that eastern Pennsylvania and the rest of the Northeast began to develop sizable residential NG markets.[8] Although NG supplied more than 60 percent of the energy required to operate households in most states by 1970, the Northeast obtained only 36 percent of its household energy from NG in that year. There are several reasons for this difference in percentages. The northeastern states are at the end of the pipeline system extending from the great producing states of Texas, Lousiana (both on and off-shore), Oklahoma, and Kansas, and the cost to deliver NG from the south central part of the United States to the Northeast is very high. These states contain few storage sites for NG and frequently must purchase such near, but expensive, substitutes for straight NG as liquefied NG (LNG) to satisfy peak winter demands for NG service. Because of the high cost of NG, the households in the Northeast that use NG use it more efficiently.

A useful way to summarize the effect of the price of NG and of other measured factors on the use of NG within households is by means of econometric analysis. Accordingly, we will use findings from econometric investigations in discussing these important effects.

ECONOMETRIC ANALYSIS

The residential demand for NG across years and states in the United States has in recent years been studied by Beierlein et al.,[9] Blattenberger et al.,[10] Grady,[11] Lin et al.[12] and Herbert.[13] An early study by Balestra[14] prompted one of the more important methodological articles in econometric literature by Balestra and Nerlove.[15] The data and equations used in many of these studies have also been evaluated, both extensively by Doman et al.[16] and selectively by Herbert,[17] Herbert and Kott,[18] and Herbert and Barber.[19] Thus we have available for this study estimated results from several studies as well as evaluations of the data used in such studies.

We will fully explore these previous studies, and we will use their findings along with findings from estimations conducted specifically for this study to examine the relationship between the price of NG and demand for NG per customer. Particular attention will be directed to the Balestra econometric analysis of NG markets since it was instrumental in all other econometric studies of energy markets that used historical data. The interpretation and reliability of estimated own-price elasticities (the effect on NG demand in percentage terms of a one-percent change in NG price) will be examined since these estimates are used widely at government agencies, utility commissions, and at NG utilities for decision making.

Results from errors-in-variables econometric analyses will be reported here. Both the nature of the data and the problem as discussed in Herbert,[20] and recent theoretical and conceptual improvements in errors-in-variables regression analysis support this emphasis. Although the importance of errors-in-variables for an econometric investigation was much discussed in early econometrics literature, such as the major early works of Frisch[21] and Koopmans,[22] it is only recently that our theoretical understanding of this earlier work has been greatly improved upon by Kalman,[23] Klepper and Leamer,[24] and Fuller.[25] Therefore, it is considered particularly appropriate to report the results of applying this methodology here.

NOTES

1. United States Department of Interior, United States Geological Survey, *Mineral Resources of the United States*, Natural Gas (Washington, D.C., U.S. Government Printing Office, 1888-1923). United States Department of Commerce, Bureau of Mines, *Minerals Yearbook*, Natural Gas (Washington, D.C.: U.S. Government Printing Office, 1924-1931). United States Department of Interior, Bureau of Mines, *Minerals Yearbook*, Natural Gas (Washington, D.C.: U.S. Government Printing Office, 1932-1978).

2. J. H. Herbert, "An Analysis of Monthly Sales of Natural Gas to Residential Customers in the United States," *Energy Systems and Policy*, 10 (1986): 127-148. J. H. Herbert, "A Data Analysis of Sales of Natural Gas to Households in the United States," *Journal of Applied Statistics*, 13 (1986): 199-211. J. H. Herbert and E. Kreil, "Supplement to 'An Analysis of Monthly Sales of Natural Gas to Residential Customers in the United States,'" *Energy Systems and Policy*, 12 (1988): 43-46. J. H. Herbert and E. Kreil, "Specification and Evaluation of Equations to Estimate Monthly Natural Gas Consumption," *Applied Economics*, 21 (1989): 1369-1382.

3. For general discussions of how labor-saving devices and the value of leisure have affected the household in the twentieth century, two texts are particularly useful: R. S. Cowan, *More Work for Mother* (New York: Basic Books Inc., 1983); Ogburn, W. F. and N. M. Nimhoff, *Technology and the Changing Family* (Cambridge, Mass.: Houghton Mifflin, 1955).

4. E. Nichols, *Public Utility Service and Discrimination* (Washington, D.C.: Public Utilities Reports, Inc., 1928), 902-1020.

5. United States Congress, House Subcommittee of the Committee on Interstate and Foreign Commerce, *Natural Gas Bill, H.R. 11662*, 74th Cong., 2nd sess., 1936, 47-70.

6. M. H. Rose, "Urban Environments and Technological Innovations: Energy Choices in Denver and Kansas City, 1900-1940," *Technology and Culture*, 25 (1981): 503-539. J. H. Herbert, "The Development of Residential Natural Gas Markets in the East North Central U.S.A. since 1940," *Energy*, 13 (1988): 393-400.

7. United States Department of Labor, Bureau of Labor Statistics, *Locally Important Fuels, Bulletin No. 950, Average Retail Price by City, 1941-1948*, (Washington, D.C.: U.S. Government Printing Office, 1949), Table 6.

8. J. H. Herbert, "Institutional and Economic Factors in Residential Gas Markets in the Northeastern United States between 1960-1984," *Energy*, 13 (1988): 211-216.

9. J. G. Beierlein, J. W. Dunn, and J. C. McConnon, "The Demand for Electricity and Natural Gas in the Northeastern United States," *The Review of Economics and Statistics*, 63 (1981): 403-408.

10. G. R. Blattenberger, L. D. Taylor, and R. K. Rennhack, "Natural Gas Availability and the Residential Demand for Energy," *The Energy Journal*, 4 (1983): 23-45.

11. S. T. Grady, "Regional Demand for Natural Gas in the Residential Sector," *Review of Regional Studies*, 16 (1986): 19-28.

12. W. T. Lin, Y. H. Chen, and R. Chatov, "The Demand for Natural Gas, Electricity, and Heating Oil in the United States," *Resources and Energy*, 9 (1987): 233-258.

13. J. H. Herbert, "Data Analysis, Specification, and Estimation of an Aggregate Relationship for Sales of Natural Gas per Customer," *Journal of Economic and Social Measurement*, 5 (1986): 165-174. J. H. Herbert, "Data Matters - Specification and Estimation of Natural Gas Demand Per Customer in the Northeastern United States," *Computational Statistics and Data Analysis*, 5 (1987): 67-78. J. H. Herbert, "Demand for Natural Gas at the State Level-Twenty Years of Effort," *Review of Regional Studies*, 17 (1987): 79-87. J. H. Herbert, "A Data Analysis of Residential Demand for Natural Gas Using Step-Down Multiple Comparison Tests," *Applied Stochastic Models and Data Analysis*, 4 (1988): 149-157.

14. P. Balestra, *The Demand for Natural Gas in the United States*, (Amsterdam: North Holland Publishing Company, 1967).

15. P. Balestra and M. Nerlove, "Pooling Cross Section and Time Series Data in the Estimation of a Dynamic Model: The Demand for Natural Gas," *Econometrica*, 34 (1966): 585-613.

16. L. A. Doman, J. H. Herbert, and R. Miller, *An Assessment of the Quality of Selected EIA Data Series - Energy Consumption Data*, (Washington, D.C.: Energy Information Administration, 1986).

17. J. H. Herbert, "Measurement Error and the Estimation of Regression Equations-A Case Study," *1987 Proceedings of the Section on Economic and Business Statistics* (Alexandria, Va.: American Statistical Association, 1988), 187-190. J. H. Herbert, "Reporting the Uncertainty in Regression Coefficients from Errors in Variables," *1988 Proceedings of the Section on Economic and Business Statistics*, (Alexandria, Va.: American Statistical Association, 1989), 259-264.

18. J. H. Herbert and P. Kott, "An Empirical Note on Regressions with and without a Poorly Measured Variable Using Gas Demand as a Case Study," *The Statistician*, 37 (1988): 387-391.

19. J. H. Herbert and L. Barber, "Regional Residential Natural Gas Demand: Some Comments," *Resources and Energy*, 10 (1988): 387-391.

20. J. H. Herbert, "A Data Analysis and Bayesian Framework for Errors- in-Variables," in *Computer Science and Statistics - Proceedings of the 20th Symposium on the Interface*, ed. E. J. Wegman, D. F. Gantz, and J. J. Miller (Alexandria, Va.: American Statistical Association, 1988), 490-499.

21. Frisch, R., *Statistical Confluence Analysis by Means of Complete Regression Systems* (Oslo, Norway: University Institute of Economics, 1934).

22. Koopmans, T., *Linear Regression Analysis of Economic Time Series* (Haarlem, Netherlands: De Erven F. Bohn N. V. 1937).

23. Kalman, R. E., "System Identification from Noisy Data," in *Dynamical Systems II*, ed. A. R. Bednarik and L. Cesari (New York: Academic Press, 1982), 135-163.

24. S. Klepper and E. E. Leamer, "Consistent Sets of Estimates for Regressions with Errors in all Variables," *Econometrica, 52* (1984): 163-183.

25. Fuller, W. A., *Measurement Error Models* (New York: John Wiley & Sons, 1988).

2

THE MARKET NEAR THE TURN
OF THE CENTURY

In this the last decade of the twentieth century, some of the largest businesses in the country uninterruptedly distribute NG to operate automatic furnaces, water heaters, clothes dryers, and other appliances within households in all our major cities upon demand. The situation was dramatically different in this century's first and second decades. In fact, many of the NG wells near the turn of the century were being operated by farmers for their own private use. Many farmers used NG to manually ignite lighting fixtures, furnaces, and other appliances. In 1910 Pennsylvania produced the most NG in the nation; 345 NG producers out of a total of 819 were producing NG for their own private use.[1]

NG was considered to be "the most useful, convenient, and luxurious fuel".[2] It was for this reason that farmers were willing to incur the expense of drilling wells and piping NG from wells to their own house even if the wells produced insufficient amounts of NG for market distribution. In addition to private use consumers, an ever-increasing number of industrial and residential (domestic) users began to purchase natural gas as markets quickly developed. In 1906, 389 billion cubic feet (Bcf) of NG was sold for $47 million.[3] A short four years later, these quantities had increased by 31 percent and 51 percent, respectively.[4] By 1970, the quantity of NG sold would be more than fifty times as great as its 1906 value.[5]

THE EXTENSIVE USE OF NATURAL GAS IN SOME HOUSEHOLDS

Domestic customers[6] used NG in ranges, lamps, furnaces, and in a variety of apparatus crafted by tinsmiths, ironmakers, and home craftsmen. The average domestic NG customer in Pennsylvania, the state with the most domestic customers in the nation in 1902, had approximately two fires, as major NG outlets within a household were called.[7] Each domestic customer in Pennsylvania used an estimated 150 thousand cubic feet (Mcf) of NG in 1906. Thus, each outlet[8] burned about 75 Mcf per year, about three times the amount of NG used by a typical customer for cooking and lighting purposes in 1975.[9] The large amount of NG used per outlet in 1902 suggests either that enormous amounts of NG were wasted within the household or that domestic customers used NG not only for lighting and cooking, as is commonly thought, but much more extensively in an assortment of makeshift and store bought appliances.[10]

Heating devices were sometimes constructed by placing a large hollow metal ring at the end of a Bunsen burner and using the ring to conduct and radiate heat within the room. A similar apparatus was attached to a burner on a range. Some utilities made such heating appliances available free of charge to their customers with a warning as to their safety. Another device was to place a large number of pans, pots, and metal shelves onto a large range and to use these metal products to conduct energy throughout the room.

The illegal means used to obtain energy free of charge were at least as intriguing as the legal, but possibly unsafe, means used to extract as much energy as possible from a NG line. In manufactured gas (MG) regions, where gas was expensive and pay-as-you-go meters were installed in many households, tales were told of customers who deposited ice wafers in the shape of quarters into these meters in order to use gas more extensively than their pocketbooks allowed.[11] Although some domestic customers managed to obtain gas for free, most paid much more than the other major customer class at the turn of the century - the industrial customer.

DIFFERENCES IN THE PRICE CHARGED DOMESTIC AND INDUSTRIAL CUSTOMERS

Even though industrial NG customers had developed more extensive uses of NG than had domestic customers, revenues generated by domestic sales were double and triple revenues from industrial sales. In large part this was because domestic customers were charged several times as much for NG as industrial customers were charged (see Table 2.1). This large price difference was frequently a controversial issue, but NG companies and government representatives often cited the higher cost of serving residential customers. A number of factors made domestic service more costly. Domestic customers were generally some distance from the NG well. Many industrial customers located manufacturing plants and mining sites near NG wells. The pressure of the NG in the pipeline needed to be reduced at regulating stations before being delivered to households. Some industrial customers could use the NG at pipeline pressure. Service lines had to be built from the distribution lines to domestic residences. Some industrial customers frequently built their own service lines to distribute NG within the plant. Domestic customers may have paid not only for the NG but also for the more extensive pipeline service and for the NG that leaked during distribution.

Table 2.1

Industrial and Domestic Consumption in 1910

	Industrial			Domestic		
	Use[a]	Cost[b]	Rate[c]	Use	Cost	Rate
Ohio	47.5	12.8	149	60.5	25.0	1.3
Pennsylvania	125.5	10.3	306	43.4	25.3	1.4
Kansas	59.5	6.9	421	23.8	21.8	1.3
West Virginia	63.4	5.1	238	13.6	17.4	1.6
New York	1.9	16.3	26	12.2	29.7	1.1
Oklahoma	21.1	4.7	136	5.4	16.9	1.4
Indiana	1.4	12.1	50	4.3	30.1	1.2
Kentucky	2.4	8.0	214	2.6	27.9	0.9
Illinois	5.4	6.1	207	1.3	21.9	1.3
Texas	6.5	6.8	142	1.6	31.8	0.6
Colorado	2.0	6.1	145	0.7	24.8	1.2
California	2.5	11.2	115	0.2	79.2	0.2
Other States	0.0	45.0	27	0.0	49.0	0.6

Source: *Mineral Resources*, 1911, 285.

Note: Kansas includes some gas piped from Kansas to Missouri and from Oklahoma to Kansas and Missouri. West Virginia includes some gas piped from West Virginia to Maryland. Indiana includes some gas piped from Indiana to Chicago, Illinois. Illinois includes some gas piped from Illinois to Vincennes, Indiana. Texas includes Louisiana and Alabama statistics. There were 320 and 6 industrial and 8,547 and 95 domestic customers in Louisiana and Alabama, respectively. Colorado includes Arkansas and Wyoming statistics. There were 121 and 4 industrial and 4,422 and 353 domestic customers in Arkansas and Wyoming, respectively. The Other States include the states of South Dakota, Missouri, North Dakota, Michigan, Tennessee, and Iowa. Total amounts of natural gas used by industrial and domestic customers in these states were 46,540,000 cubic feet and 64,098,000 cubic feet, respectively.

[a] Use is consumption in billion cubic feet.

[b] Cost is price in cents per thousand cubic feet.

[c] Rate is use per 10,000 customers in billion cubic feet.

The relatively low price charged industrial customers for NG was justified, in part, because the cost of serving these customer was less. Moreover, contracts to serve NG to large industrial customers, because of the large amounts of NG they used, could be used to justify the extension of a pipeline system into a region.

Industrial customers were large customers with market power. The average industrial customer consumed much more NG than the average residential customer (see Table 2.1). Some large industrial users of NG, such as carbon black plants, had the capability of relocating to regions where the NG was least expensive. Relatively inexpensive coal was also readily available in regions where NG was being marketed to industrial customers. If NG producers or distributors wanted to sell NG to some industrial customers, they had to be willing to offer it at a price that was competitive with the industrial price of coal and competitive with the lowest cost producer of NG in a region. Even though there were some NG customers for whom NG was a premium fuel, such as glass manufacturers, there were many more industrial customers for whom NG was just one more possible source of energy to operate and heat plants.[12]

EARLY MAJOR NATURAL GAS GROWTH MARKETS

A major proportion of NG sales to industrial customers was to manufacturing establishments in western Pennsylvania, northern West Virginia, and eastern Ohio. This part of the country had both abundant energy and labor, which kept down the cost of both factors of production, and manufacturers who were capable of putting these resources to work. These manufacturers used the relatively inexpensive energy and labor to produce the steel, glassware, and other products required for the growing American economy and for its growing export markets.

Between 1906 and 1910, 4,962,000 immigrants entered the country, the highest five-year total ever in U.S. history.[13] Many of these immigrants settled in western Pennsylvania, West Virginia, and eastern Ohio to work in the factories and mines. This region due to the combination of cheap gas and labor, plus accessibilty to iron and other raw materials developed into the major NG consuming region of the world.

Because of its booming industry, life improved considerably for many citizens in this area. Income and employment were growing. Between January 1904 and January 1914, the number of workers employed in manufacturing increased by 84 percent, 46 percent, and 27 percent, in West Virginia, Ohio, and Pennsylvania respectively, and the average wage of these workers increased by as much as 61 percent in Ohio. Growth was especially rapid in such cities as Akron, Ohio. The manufacturing labor force increased from 9,621 in 1904 to 24,680 in 1914 in Akron.[14] Even the quality of household products improved. The special properties of NG, for example, meant that glass and metal products for the home could be refined and improved, leading in turn to an increase in demand throughout the United States and abroad.

The NG industry itself required both a significant labor force and many miles of iron pipe as it inched out across towns, cities, and rural areas of Pennsylvania, Ohio, and West Virginia. Large teams of men, mules, and horses were used to dig ditches, lay pipes, and set huge compressors. These teams and their leaders were legendary and could be observed working year round in different parts of this region.[15] Daniel O'Day, who worked for John D. Rockefeller at National Transit Company, the major gas pipeline company at the time, and who was known to be quick to defend his immigrant Catholic workers from local bigotry, was one of these legends. "No terrain was too difficult for an O'Day pipeline. No winter too cold or ground too hard. If you didn't want to work you didn't work for Daniel O'Day."[16]

NATURAL GAS REGIONAL MARKETS

By 1910 domestic NG customers could be found in twenty two states.[17] Yet four states had fewer than 100 customers, and another four states had fewer than 1,000 customers.

Although Texas, California, and Louisiania had more than 8,000 customers each, they were primarily oil states. California alone produced 28 percent of the total value of petroleum in the United States (see Table 2.2). In these states NG production was frequently incidental to the production of oil and was frequently viewed as a nuisance rather than as a valuable product. The value of oil produced in these states was several times as large as the value of natural gas. On the other hand, the value of NG produced in West Virginia and Ohio was much greater than the value of petroleum.

Of particular interest are those states where NG was clearly a valued commodity, where many domestic NG customers lived, and where many cities, towns, villages, and farms were served with NG. States that met these criteria fell into three regional RNG markets in the first two decades of the twentieth century - an old and growing northern market, a declining midcontinent market and a new and growing south central market.

Northern Market

West Virginia, Pennsylvania, Ohio, and New York all shared three essential ingredients that led to the development of major RNG markets in each state -- rich natural resources, high population density[18] (which meant both an abundant labor force and high demand), and nearby industries that both produced the pipe for transporting NG and depended on the NG produced for energy. A high population density allowed the pipeline industry to continually build relatively short pipeline extensions in order to serve one more town or village. In 1912, 443 cities, towns and villages in Ohio were being served NG, yet by 1917 the number of places served NG had increased to 529.[19] Many of these communities were small but the expectation of the company managers was that the economy of states such as Ohio

Table 2.2

Value of Natural Gas and Petroleum Produced in 1910

	Gas	Petroleum	Total
West Virginia	$23,816,553	$15,720,184	$39,536,737
Pennsylvania	21,057,211	11,908,914	32,966,125
Ohio	8,626,954	10,651,568	19,278,522
Kansas	7,755,367	447,763	8,200,130
Oklahoma	3,490,704	19,922,660	23,413,364
Louisiana	956,683	3,574,069	11,136,507
Illinois	613,642	19,669,383	20,283,025
California	476,697	35,749,473	36,226,170
Texas		6,605,755	
Other*	3,962,347	3,649,559	695,663

Source: *Mineral Resources*, 1911, 288.

* Includes the states of Arkansas, Colorado, Wyoming, Utah, South Dakota, Missouri, Michigan, North Dakota, Tennessee, Iowa, New York, Kentucky, and Indiana. In both New York and Indiana the value of Natural Gas and Petroleum was about $1,500,000. Value of Natural Gas includes some gas from Texas and Alabama. Value of Petroleum includes some petroleum from Alabama. Combined Value includes some natural gas and petroleum from Alabama and Texas.

would continue to grow, and thus these markets would become increasingly lucrative. Managers of the companies also thought that as customers in these places became accustomed to the convenience of NG they would continue to use NG even when the price rose. Major development was confined largely to the western portions of Pennsylvania and West Virginia, the southwestern portion of New York, and the eastern portion of Ohio. The northern market was an interstate market, with several pipelines over 100 miles in length and with interconnected pipelines which extended over wide areas. Out of 24,973 miles of NG gas pipeline in the United States in 1902, 37 percent, 15 percent, 11 percent, and 5 percent were in Pennsylvania, Ohio, West Virginia, and New York, respectively.[20] In particular, there was an extensive pipeline system around the Pittsburgh area.

Households and businesses in Pennsylvania, Ohio, and New York consumed much more NG than those states produced (see Table 2.3). West Virginia, which produced 37 percent of the NG produced in the United States in 1910, shipped NG to Pennsylvania and Ohio. And Pennsylvania, which couldn't meet its own demand for NG, supplied NG to New York, the result of New York consumers' willingness to pay one and one-half times what their Pennsylvania counterparts paid for the commodity. Eventually, the northern regional market expanded to include Maryland and Kentucky, as NG was piped to these states from West Virginia. However, the states of Maryland and Kentucky did not become significant RNG markets until later in the century. By 1913 Indiana joined the northern market, as pipelines were connected from Harrison and Lewis counties in West Virginia to Sugar Grove, Ohio, and then onto Muncie, Indiana.[21] This important connection represented an attempt to shore up the Indiana NG market, which had declined over the years.

Table 2.3

Production and Consumption in Northern Regional Market in 1910

	Production		Consumption	
State	Volume[a]	Price[b]	Volume	Price
West Virginia	190.7	12.49	77.1	7.16
Pennsylvania	126.9	16.60	168.9	14.17
Ohio	48.2	17.89	108.1	19.64
New York	6.0	27.93	14.2	27.92

Source: *Mineral Resources*, 1911, 284.

[a] In billion cubic feet of natural gas.

[b] In cents per thousand cubic feet at the point of consumption for both production and consumption amounts and not at the well-head for production amounts.

Midcontinent Market

Near the turn of the twentieth century, a significant RNG market existed within Indiana, along with some trade between Illinois and Indiana. In fact, in 1897 with 214,750 domestic fires, Indiana had more domestic fires than any other state, even Ohio, which had only 85,638 fires.[22] But by 1910 Ohio had 475,505 domestic customers, more than any other state, whereas Indiana had only 36,054 domestic customers.[23] The market in the midcontinent region declined significantly throughout the first decade of the twentieth century and was the first clear instance of a NG market failing due to lack of planning and regulation. Enormous amounts of NG were wasted both at the production site and in the transmission of NG from the site. As many NG wells began to run dry, producers returned their property rights in wells to the farmers who owned the land. In most cases the farmers gladly accepted and became producers, since the wells usually held enough NG for the farmers' private use.

In this way, while the number of domestic and industrial consumers of NG and the number of producing wells in Indiana declined between 1906 and 1910, the number of producers almost doubled. In fact, in 1918 the number of producers, many of whom were farmers, were at 1902 levels, yet the number of producing wells had declined by more than 66 percent (see Table 2.4). Thus, when the rest of the country was going from private to public use, Indiana was going from public to private.

Table 2.4

Natural Gas Market Declines in Indiana around 1900

	Domestic	Industrial	Producers	Wells
1902	101,481	1282	929	5,820
1910	36,054	282	1027	2,955
1918	31,023	284	935	1,760

Source: *Mineral Resources*, 1917, 1084. *Mineral Resources*, 1918, 1397, 1401.

Note: Amounts given are actual numbers of domestic consumers, industrial consumers, producers and producing wells.

South-Central Market

The 1910 RNG market was particularly well developed in Wichita, Kansas, and Kansas City, Missouri[24]. The local utility in Kansas City aggressively recruited domestic customers and emphasized the improved efficiency and cleanliness of NG compared to coal and wood. Some of this NG was supplied from wells in Oklahoma. By 1910 Kansas, Oklahoma, and western Missouri made up a thriving south-central regional market (see Table 2.5). Oklahoma became a major producing state by 1910 because of a Supreme Court ruling. Formerly Indian territory, Oklahoma received statehood at the close of 1907. A provision in its constitution restricted transportation of NG outside the state. The Attorney General of Oklahoma argued that the provision did not interfere with commerce since Oklahoma had the right to conserve the supply of NG for its citizens. The United States Supreme Court overturned the provision and argued that it interfered with interstate commerce.[25] As a result of this ruling, NG customers in Kansas and Missouri could be served NG from Oklahoma, and the value of NG production in Oklahoma soared from $860,159 in 1908 to $3,490,704 in 1910.

Although for a period of time in Kansas NG was a widely available source of energy for operating household appliances, Kansas customers would soon experience curtailments in service and other restrictions on NG use as several NG fields began to run dry. In 1911 the Kansas board of public utilities created a commission headed by Erasmus Haworth, State Geologist, to assess the state's NG supply.[26] Based on an assessment of the productive capacity of the known fields and those likely to be discovered, Haworth concluded that the state had only enough NG to last for three or four more years.

Table 2.5

Production and Consumption in South Central Regional Market in 1910

	Production		Consumption	
	Volume[a]	Price[b]	Volume	Price
Kansas	59.4	13.06	81.9	11.19
Oklahoma	50.4	6.92	27.9	7.47
United States	509.2	13.90	509.2	13.90

Source: *Mineral Resources*, 1911, 284.
Note: Kansas consumption includes gas piped from Kansas and consumed in Missouri and gas piped from Oklahoma into Kansas. Oklahoma consumption includes some natural gas which was piped to Missouri.
[a] In billion cubic feet of natural gas.
[b] In cents per thousand cubic feet at the point of consumption for both consumption and production amounts and not at the well-head for production amounts.

Until then, some domestic customers in Kansas and elsewhere were still charged a flat rate for NG service. Some considered that flat rates were the major cause of the supply problems in Kansas.

FLAT RATES

Flat rates were used at times in places where NG was plentiful. Flat rates were also used in some places where "customers using natural gas felt that it was a gift of nature and objections were raised when rates above a nominal amount were charged."[27]

The flat rate could take the form of a fixed dollar payment per month for NG service or a fixed dollar payment per appliance. Such rates were common in the nineteenth century for both electric and gas service. Flat rates were used to encourage households in a particular area to receive electric and gas service for the first time. Moreover, if the average use per customer was predictable and was relatively constant throughout the year, the particular utility serving the area could even improve the opportunites for scale economies by offering flat rates.

Flat rates were also common in the early twentieth century for phone service. Such charges also encouraged households to become phone customers. The more households that used phones the greater the value of phone service to each individual phone customer, because more households were accessible by phone. However, the flat rate had to be fixed high enough to cover the increased cost of the increased level of phone service since scale economies were not to be obtained in the telephone service industry. The unit cost of phone service tended to rise as the level of phone activity rose.

Flat rates were used for a longer period of time for water service than for any other utility service, since the cost of a meter and a meter reader were prohibitively expensive, given the cost of serving water to residents of a community.

Flat rate fees were used to encourage large numbers of households to hook up to a gas line and, thus, flat rates improved the prospects for long-term profits. However, as concern for the depletion of NG wells mounted, flat rate fees became controversial and were considered to lead to great waste and inefficiency. Eventually, the United States Department of the Interior, in the case of Kansas specifically, recommended the abolition of the flat rate in the 1911 edition of the Mineral Resources of the United States.

As the technical and economic features of meters were improved, and as it became generally understood that NG was an exhaustible resource, flat rates were being replaced by a fixed charge for each cubic foot of gas sold. These rates were known as straight-line meter rates. These rates, in turn, were being replaced by block rates in which the cost of NG declined as larger amounts (blocks) of NG were used. However, block rates were much more common for electric service than for gas service because economies of scale were much greater for the generation of electrical energy than for the distribution of NG.[28]

After 1916, the RNG market in Kansas was very unstable (see Table 2.6). Flat

Table 2.6

Domestic Market in Kansas between 1910 and 1919

	1916	1917	1918	1919	Average[f]
Sales[a]	21	9	15	12	23(3.0)
Cost[b]	22	43	19	20	23(0.7)
Customers[c]	202	188	120	116	194(5.6)
Average Sales[d]	103	48	120	104	117(14)
Deflator[e]	.17	.37	.61	.85	.05(.02)

Sources: Sales, Cost and Customers, *Mineral Resources*, 1910-1921, various pages; Deflator, Consumer Price Index-All Items, *Historical Statistics, Census*, 211.

[a] In billion cubic feet.

[b] In cents per thousand cubic feet in constant dollars.

[c] In thousands of customers.

[d] Sales per customer in thousand cubic feet.

[e] The reported deflator is the Consumer Price Index expressed as a proportionate change from its 1910 value. By 1919 prices were 85% higher than their 1910 level.

[f] Average for the years 1910 to 1915. Standard errors are given in parenthesis.

rates were one explanation for the problems in Kansas. Another explanation was supply problems in obtaining natural gas from Oklahoma.

Similar but more isolated system failures occurred in other parts of the country as well, as poor planning and unexpected depletion of particular gas fields led to gas shortages. Many of these problems would eventually be solved by piping NG from producing areas that had excess NG and injecting this gas underground in depleted wells near market areas in the summer for distribution to domestic customers during the late fall and winter months.

THE IMPORTANCE OF UNDERGROUND STORAGE

Few underground storage reservoirs existed in the United States before 1936. The Zoar storage facility built in New York in 1916 was the first underground storage reservoir. It was followed by the Menifee reservoir in Kentucky in 1919 and the Queen reservoir in Pennsylvania in 1920. Gas storage reservoirs or holders above ground were more common. These looked much like oil or water tanks and had limited storage capacity.

The gas storage reservoir at Menifee[29] was developed under the simple assumption that just as NG was stored and trapped in the earth for millions of years before being withdrawn by man, it could be stored again in the same earth. As gas was withdrawn from the site at Menifee, the pressure of the site became so low that space heating needs of nearby domestic customers were unable to be adequately served during the winter. After pumping NG back into the ground, the pressure of the site rose and space heating needs were once more able to be satisfied from the wells at Menifee.

In general, the lack of storage facilities constrained the development of some RNG markets. Without adequate underground NG storage or wells with sufficient pressure, the NG industry found it difficult to economically satisfy the peak winter demands for NG of many space heating customers; demands that were several times larger than average summertime demands. There were instances where the NG industry needed to construct pipeline systems of sufficient capacity for the entire distance from NG wells to distribution points near domestic space heating customers in order to adequately serve wintertime demands. The system was operated at full capacity during the winter and at much lower capacity during the summer. If storage sites were available near domestic customers, however, then only that portion of the pipeline system, from the storage site to the domestic customer, would need to be of sufficient capacity to serve domestic customers during the winter.

Although underground storage additions were few, extensions to pipeline systems were constantly being made near the turn of twentieth century to effectively serve markets and these pipeline systems were becoming increasingly interconnected both physically and financially.

INTERCONNECTED PIPELINES

West Virginia was the major source of supply for NG in the United States. An extensive interconnected pipeline system extended outwards from West Virginia; more than 50 percent of the NG produced in West Virginia was delivered to consumers in contiguous states. NG from West Virginia was regularly shipped to Pennsylvania, Ohio, Kentucky, and Maryland, and NG flowed from these states to West Virginia, but in much smaller quantities.

The NG used in households in Cleveland and in other cities on the south shore of Lake Erie came from West Virginia. By 1903 an 18-inch pipeline 118 miles long, extended from the Ohio River to Cleveland, and by 1909 a 20-inch pipeline extended from West Virginia across the northern border of Kentucky to Cincinnati, Ohio. By 1920 interconnecting pipeline systems formed a network over the entire region from the Allegheny mountains west beyond the Ohio-Indiana border and north to Lake Erie. According to market experts, in 1922 these pipelines formed "one of the most elaborate networks yet for the transfer of energy by engineering science. They cover an area almost as wide and deal with quantities of potential energies almost as great as those considered in the proposed Boston to Washington 'superpower system' of interconnected electric-service facilities".[30]

Because transportation of NG was relatively expensive and there was no direct control over interstate transactions in NG, the cost of NG was much greater in a state that relied on imported NG, such as Ohio, than in a net producing state, such as West Virginia. For example, the typical NG customer living in Ohio would have to pay 75 percent more for the NG produced in West Virginia than would the typical customer living in West Virginia.

While the West Virginia legislature was pleased with income brought into the state by the export of NG, it was concerned that sufficient supplies of NG might not be available at all times to serve the needs of its citizens. The legislature enacted statutes requiring pipelines serving West Virginia customers to give priority to those customers over customers residing in other states. The statute met with great opposition from consumers in Pennsylvania and Ohio, and its constitutionality was considered by the United States Supreme Court in the latter part of the decade.[31]

The Supreme Court eventually ruled that the statute "would work a practical cessation of the interstate stream and that the law was a 'prohibited interference' with interstate commerce." Justice Holmes, however, dissented since he saw nothing to prevent a state from giving preference to its own citizens in the enjoyment of its natural resources, especially if the gas were used for private purposes. Holmes knew "of no power in Congress to require them to devote it to public use or to transport it across state lines. It is the law of West Virginia and of West Virginia alone that makes the West Virginia gas what is called a public utility, and how far it shall be such is a matter that the law alone decides."[32]

By 1920 nearly 15 percent of the NG produced in the United States was being transported among states. The interconnected pipeline system enabled distribution companies to obtain NG from several sources and avoid NG service curtailments when a local NG reservoir ran dry.[33]

Interconnected pipeline systems were, in part, a consequence of the consolidation of independent gas companies into single corporate entities. Such consolidations were a major movement during the second decade of the century. Company consolidations resulted in reduction in the irrational production behavior motivated by the law of capture, which allowed any company operating in an area to produce as much NG as it desired. As a result of each company pursuing its own self-interest, gas fields were prematurely depleted and much NG was wasted. By consolidating the joint company could regulate the production of NG and increase long-run supplies. However, consolidation also tended to reduce the responsiveness of these businesses either to consumer desires or to current economic conditions. Additions to pipeline systems after consolidation were frequently undertaken to gain control of areas for long term strategic objectives of a large corporation and were not undertaken to supply a superior type of energy service for the household.

Company consolidation could also increase the size of the NG market if the company acquired by the NG company was a MG company. Many MG companies were converted to NG distribution companies as a consequence of consolidation. This conversion varied from being relatively easy to difficult, depending on differences in the physical characteristics of the MG and the NG.

USE OF MANUFACTURED GAS

MG, as well as NG, was used widely by domestic customers during the first several decades of the twentieth century.[34] In 1912, 158 Bcf of MG and 193 Bcf of NG was sold to domestic customers.[35] In 1920, the amount sold had grown to 263 Bcf of MG and 286 Bcf of NG.[36] The widespread use of expensive MG attests to the high value of energy in a gaseous form rather than in a liquid or solid form. Gas was manufactured in a wide variety of processes from coal, oil, animal fats, and any other item containing hydrocarbons.

Many of the major NG utilities of the twentieth century were originally gas manufacturing plants. They developed their gas distribution network to transport MG, not NG. MG contained less usable energy per cubic foot than NG and was generally much more expensive than other sources of available energy. However, just like NG, it also supplied energy to households in a form that could instantly provide light and heat.

MG was widely used as an illuminant before the turn of the century. It was still widely used for that purpose as late as 1910, even though electric lighting had been available in many places for thirty years. Gas light had many supporters.

Robert Louis Stevenson compared electricity with gas as a source of street lighting in the following way: "Such a light as this should shine forth only on murders and public crimes, or along the corridors of lunatic asylums, a horror to heighten horror. To look at it only once is to fall in love with gas, which gives a warm domestic radiance fit to eat by."[37]

As electricity became more available and less expensive, and the quality of electric lighting improved, gas lights were replaced. Instead, gas was increasingly being used as a heating and cooking fuel[38] rather than as an illuminant. The competition for the lighting market was an early example of the competition between gas and electricity in supplying energy for the household. The transition from gas light to electric light accelerated during the first decade of the century.

The popularity of MG as a fuel stemmed from its convenience and its freedom from smoke and ashes when compared to coal and wood. When compared with fuel oil, the relative advantage of gas was primarily a function of price and superiority of gas as a source of heat. Yet fuel oil was preferred by some who considered it a safer source of energy. But consumers in many markets had no choice. In many instances either gas or oil was available, but not both, and consumers used what was available.

CONVENIENCE AND DANGER OF GAS USE IN HOUSEHOLD EQUIPMENT[39]

Gas was particularly convenient for household use when compared with coal and wood, the primary alternative fuels at the time. Once a house was hooked up to a main pipeline in the road, gas was distributed to a ring or rod in a makeshift or store-bought appliance or furnace. The rod or ring had outlets from which the gas could escape when a valve was opened by the turn of a handle. When the escaping gas was ignited with a match, instant heat and light were available from each of the outlets.

The speed and ease at which gas was available could have dire consequences. Severe damage could result from explosion or fire if gas was allowed to escape ffor too long a period of time from a gas appliance before a match was ignited. A spark released in switching on an electric light could also ignite the gas and set off an explosion. Yet despite the inherent dangers, gas appliances and stoves were greatly appreciated by many.

The Appearance of Gas Equipment in the Household

Ornate iron gas stoves, ranges, and room heaters were proudly displayed in many households near the turn of the century. Stoves were decorated either with tile or with an American eagle or with a French fleur-de-lis or with the name of the company that manufactured the stove or with filigree borrowed from the designwork of some cathedral. Room heaters, or radiators (if central heating was available) were also ornate, especially the legs, which rested on the floor. Although gas stoves generally took up much less space than coal and wood stoves, they were not yet automatic. The iron of a stove also retained the heat from the gas flame and provided heat to a room long after the flame was shutoff. Some stoves were also connected to a water heater, which made for a very efficient arrangement.

Gas irons, toasters, small nursery heaters, hot plates, and even artificial logs heated with gas appeared for sale in stores. Although there were mechanical problems with these appliances and improper use could result in injury and even death, their functional and aesthetic characteristics were always appreciated.

Types of Gas Equipment in the Household

Lighting fixtures were the most common household appliances throughout the nineteenth century. The gas flame from a lighting fixture was yellow and shaped like a fan until the midddle of the nineteenth century when the Bunsen burner was developed. The Bunsen burner used a mixture of air and gas to improve combustion. The flame was adjustable, and the now familiar blue gas flame appeared.

Stoves marketed during the nineteenth century were used for cooking and general heating purposes around the household. For example, the large amounts of NG used per residential customer in states such as West Virginia and Oklahoma, where NG was relatively cheap and where the loss of NG in transporting it short distances was relatively small, was simply too great to be attributed to lighting and cooking needs exclusively. For example, ranges were reported to use roughly 20 Mcf of NG annually,[40] yet the average RNG customer in 1915 in Oklahoma and West Virginia purchased 148 Mcf and 154 Mcf of NG, respectively. Thus, customers in West Virginia and Ohio probably used gas wastefully, but they clearly obtained more heat from NG within the household than was necessary to heat food. They used this energy to heat water and interior spaces.

Between 1910 and 1920 the gas industry began to market stoves more aggressively as it continued to lose lighting customers to electricity. The National Commercial Gas Association declared April 26 to May 1, 1915 Gas Range Week and published a half-page advertisement in the Saturday Evening Post and the Literary Digest. The first stove with a thermostat was marketed in 1915.

Some freestanding water heaters were sold, but many amounted to little more than a huge pot on top of a stove. In fact, hot water was still most frequently obtained by heating a large pan of water on top of a stove and carrying, rather than piping, the heated water to where it was needed. Despite technical improvements in their operational performance, systems which piped the heated water from freestanding water heaters had a reputation for being very inefficient, and few were sold.

Although many gas space heaters were marketed, most of the heavy, cast-iron central space heating units that were sold used coal since coal was the least expensive source of energy at the time. In general, central space heating units were quite expensive to purchase and operate and were generally only used in upper income households.

Central space heating gas units, when purchased, were generally placed in the basement. Heat was generated from these units and was distributed by steam, water, or mechanical means to other parts of the household. Steam and water systems worked better but required maintenance. To minimize maintenance costs, simple mechanical systems, which attempted to take advantage of the fact that heat rises, were frequently put in place. These systems directed the heat through pipes to different parts of the household. However, sufficient heat was frequently unavailable to heat the second floor of a house. In fact, the heat was frequently insufficient to properly heat a single room, and the heat that was available on a particularly cold night tended to rise to the ceiling and heat the upper portions of rooms.

For many families a wood, kerosene, or gas stove may have been the sole source of energy for the room or two that they occupied. This arrangement had some advantages. It enabled the household to avoid the initial expense plus maintenance costs of a boiler or a central space heating furnace. It also allowed NG consumers to readily reduce their use of NG by shutting off the stove with the turn of a knob when either the price of NG or the ambient temperature rose.

GAS USE PER CUSTOMER, PRICE AND WEATHER

When the average price of NG fell between 1914 and 1918 (see Table 2.7), in some key states,[41] average use per customer generally rose (see Appendix IV for more detail). However, part of the large increase in use between 1916 and 1917 may have been due to the much colder weather in 1917. Moreover, differences in the availability of NG within all parts of a state may have played a role in determining differences in the average level of use between states. Ohio, for example, had experienced deliverability problems in different parts of the state during the time period. These problems were reported in the press, which may have encouraged consumers throughout the state to use NG less extensively than they would have otherwise. West Virginia, on the other hand, had plentiful supplies of NG available within the state, and this was well publicized. Thus, average use per customer was 50 percent higher in West Virginia than in Ohio even though there was little difference in the coldness of the winter in these two states.

Table 2.7

Average Price and Consumption per Customer for Key States

	1910	1911	1912	1923	1914	1915	1916	1917	1918	1919
QD[a]	135	126	135	123	123	126	125	135	139	121
PG[b]	23	24	23	23	23	23	21	19	17	21

Sources: *Mineral Resources*, 1910-1919, various pages; Deflator, Consumer Price Index-All Items, *Historical Statistics, Census*, 211. Note: Amounts are rounded to nearest whole number.

[a] Average price per thousand cubic feet in constant dollars.
[b] Average consumption per customer in thousands of cubic feet.

In order to summarize the relationship between NG use per customer, price, availability, and changes in temperature between years (DT),[42] a relationship was estimated by least squares techniques.[43] State effects for Ohio and West Virginia were also estimated to obtain an indication of the effect of availability on NG use per customer in these two states (see Table 2.8).

Table 2.8

Estimated Effects

	Price	Temperature	State Effect West Virginia	Ohio
Effects	-1.72	-74	+21Mcf	-16Mcf
Standard Error	.32	25	4	4

It was estimated that, given a fixed temperature, for each penny increase in the average price of NG, average NG use per customer declined by 1,720 cubic feet. Looking at it another way, for each 10 percent increase in the real price of NG, NG use per customer declined by 2.9 percent, that is the price elasticity was equal to -0.29.[44]

The negative sign on the DT effect indicates that, given a fixed price, when the temperature rose between years, demand did, in fact, decline. After accounting for the effect of price and temperature on use per customer, the average use per customer was estimated to be 21,000 cubic feet greater and 16,000 cubic feet less in West Virginia and Ohio, respectively, than in the other states.[45]

Since the variables used in the estimation probably contain significant measurement error, a range for the price elasticity, which takes into account the influence of measurement error in the variables,[46] on the estimated effects, was also estimated. The estimated range was -0.29 to -0.76. Thus the price elasticity was still observed to be greater than -1.00 even when the elasticity was estimated by this more general technique.

The finding that the average domestic price elasticity was greater than negative one but less than zero helps to explain the differential price charged residential and industrial customers discussed previously. The negative sign for the price elasticity reveals that, at a fixed temperature change between years, the amount of use per customer declined, on average, when the price domestic customers paid for NG increased. However, the magnitude of the coefficient indicates that the gain in revenues from selling each unit of gas at a higher price was greater than the loss in revenues from selling a smaller volume of gas. Therefore, profits of NG companies could be increased by increasing the price to residential customers.[47] Moreover, if the NG price elasticity for industrial customers was less than -1.00, profits of NG companies could be increased by reducing the price of NG to these customers.

During this time period carbon black and gasoline manufacturers often relocated to regions where NG was plentiful and cheap. Other industrial customers produced NG for their own use, switched between fuels in operating boilers or had several sources of supply for NG. These customers were able either to reduce their use of NG or to substitute an alternative fuel for NG or to find an alternative source for NG if the price they were charged from a current supplier was to increase. Therefore, the average price elasticity for some industrial customers may have been less than -1.00 in value. Thus, another reason why industrial customers paid less for natural gas than residential customers was that it was in the short-term economic interest of many gas companies to follow a strategy of lowering the price of NG to industrial customers and increasing the price to residential customers. If NG businesses did not lower prices, there was always the chance that the industrial customer would decide to use more coal instead of NG. While the importance and availability of NG had grown during the second decade of the century,[48] coal was by far the dominant fuel used in the American economy, and it was readily available under normal conditions in most places.

NG in 1910 was a relatively small proportion of the value of energy used in the American economy (see Fig. 2.1). However, one of the reasons for the very large increase in the value of coal and petroleum prices after 1916 was the large increase in the price of these fuels, and not the large increase in the volume of shipments. While the price of coal and oil rose dramatically, the cost of NG remained relatively constant. A large proportion of the rise in the price of oil and coal stemmed from allocation problems associated with entrance into and exit out of the First World War by the United States. The federal government played a dominant role in the allocation of oil and coal, which were viewed as much more critical fuels for the operation of the wartime economy than gas. The attempts at control by the federal government and the industrial response to these controls frequently caused major distribution problems. As a consequence of such problems

Figure 2.1

Value of Mineral Fuels in the United States between 1906 and 1921

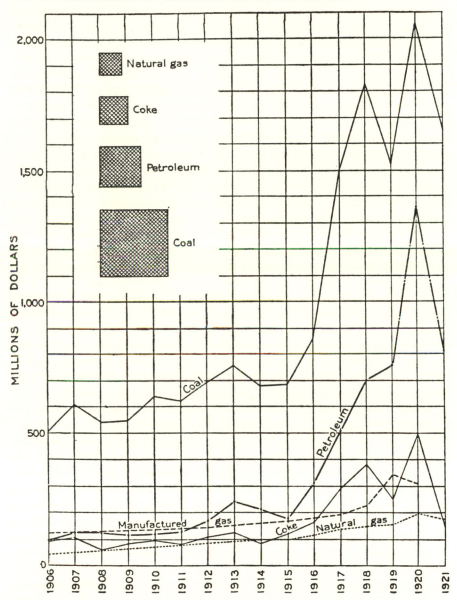

Source: *Mineral Resources*, 1921, 368

the price of these fuels rose. NG prices also rose by less than the price of the other fuels because natural gas exchange relationships were, to some extent, governed by long term contracts and were under the control of municipal and state utility commissions.

MUNICIPAL AND STATE COMMISSIONS FOR OVERSEEING GAS EXCHANGES

Municipal commissions or boards for the governance of gas exchanges were formed in many cities during the nineteenth century. State commissions for the regulation of gas exchanges generally developed only after state commissions were formed for the regulation of railroads, the natural monopoly of greatest public concern during the nineteenth century. The Massachusetts state commission for the regulation of railroads, founded in 1869, probably the most important regulatory pioneer of the nineteenth century,[49] preceded the State Board of Gas Commissioners by sixteen years. Two years later in 1887, the Board was expanded to become the Board of Gas & Electric Light Commissioners.

The Massachusetts Commission and many other state commissions of the nineteenth century did not have direct control over rates. The commissions attempted to influence rates by persuasion and by publicity.

Municipalities also attempted to influence rates when assigning franchise rights to operate gas utilities. Some cities even had direct control over rates.[50] However, most of these agencies were limited in their effective control over NG exchanges because NG exchanges extended beyond city borders. Many of these agencies were also corrupt and were using their granting of rights and control over rates largely for personal and political gain. As a consequence state utility commissioners with broad powers and direct control over rates were increasingly being established in the first several decades of the twentieth century.[51]

Although many state utility commissions founded in the twentieth century did have the right to investigate the reasonableness of rates that utilities charged their different customers for gas, they did not generally have the resources necessary to conduct such investigations. The resources that were available to the commission were used to develop administrative procedures and standards. They needed procedures for granting privileges to particular companies for serving an area. They needed standards for general accounting purposes and for determining the value of utility property for rate-making purposes. They needed standards for the heating value, pressure, and purity of the gas to be served these customers. Then they may have spent some of the remaining resources on making sure that these standards were followed. Thus the relatively modest increase in the price of gas, when compared to other types of energy after the First World War, was to some extent a consequence of a much more extensive local and state governmental oversight of gas exchanges, but was to a much greater extent probably due to few

Figure 2.2

Natural-Gas Gasoline Produced in the Nine Leading Producing States

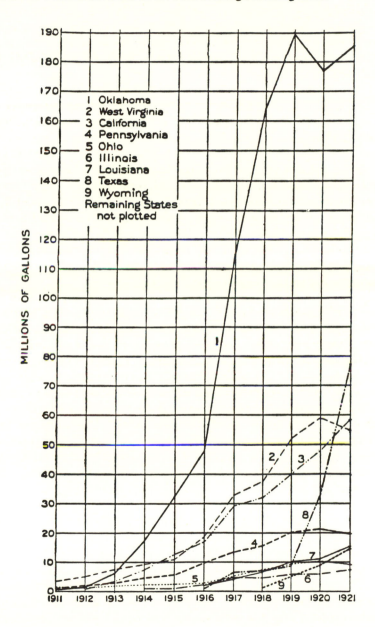

Source: *Mineral Resources*, 1920, 289

labor problems in the gas industry and to a relatively modest increase in demand for this fuel over other fuels, after the First World War, when government restrictions on the use of fuels ended.

Although government agencies may have had some control over the cost of NG to NG customers, they had little control over the direction for the development of NG use in the economy. In fact, the development of NG markets could be greatly affected by the growth of particular industries such as the previously mentioned natural-gas gasoline industry.

NATURAL GAS AND THE GROWTH OF THE NATURAL-GAS GASOLINE INDUSTRY

During the second decade of the twentieth century, NG was being processed to support the major growth sector in the United States economy at the time; the use of automobiles, trucks, and other gasoline powered vehicles. In 1920, 496 Bcf of NG was processed for its gasoline content while 798 Bcf of NG was produced for NG markets.[52] Some of the heat value and the volume of the NG was lost in this processing. However, processing of NG for its gasoline content was viewed by the National Bureau of Standards, which provided much technical support to the gas industry at the time, as enhancing the value of the NG served.[53] The processing of NG resulted in a reduction in the condensate in the NG. This condensate could result in problems with pipeline couplings, which in turn could result in leakages of NG from the pipeline. The condensate could also freeze in the pipeline during cold weather and could thus reduce carrying capacity of the pipeline when carrying capacity was most needed.

In 1920, 7 percent of the gasoline produced in the United States was obtained as a consequence of NG processing.[54] The great growth in gasoline production from a NG source (see Figure 2.2) was due to several factors. The natural-gas gasoline plants could locate near the NG production site. The difference in the cost of NG to the processing plant and the value of the gasoline produced was large, suggesting that significant profits were to be gained from this activity. For example, in 1920 the difference in value of gasoline and the cost of NG per gallon of gasoline was $0.176.[55]

SUMMARY AND CONCLUSION

The RNG market in 1910 offered many contrasts. The industry was not well organized. The amount of NG sold in most parts of the country was small. Flat fees were still used in some places, safety in use was sometimes lacking, and waste was common. Yet, a sizable market and an interstate commerce had developed in several parts of the country: a declining midcontinent region in

Indiana, a minor south central region comprised of Oklahoma and Kansas, and a major northern region made up of parts of West Virginia, Ohio, Pennsylvania, and New York. NG was being transported from production sites, where its value was low, to markets, where its value was greater. Public bodies were formed to supply information about and regulations for this exhaustible resource. Moreover, the level of use per customer between 1910 and 1919 would decline after this decade and would not reach levels observed for this decade until the 1970's in many key energy-using states.

Estimated results presented here indicate that the own-price elasticity for NG demand per domestic customer was clearly negative but still greater than -1.00 in magnitude. Thus, a percentage increase in the price of NG, other things being equal, resulted in an increase in the revenues received by gas companies. It was in the interest of the industry to increase the price of RNG in a region where they were already supplying NG to a large proportion of the households. Finally, the aggregate indirect contribution of NG to the welfare of RNG customers, through increased employment and reduced prices for other goods and services purchased by these customers, may have been greater than the direct contribution through a cleaner and more efficient household requiring much less labor to support energy intensive activities within the household.

NOTES

1. *Mineral Resources*, 1910, 309. After 1914 the proportion of private producers to total producers tended to decline. By 1914 there were 401 and 1,561 private producers out of 1,325 and 2,268 total producers in Pennsylvania and Ohio, respectively.

2. *Mineral Resources*, 1908, 317.

3. *Mineral Resources*, 1906, 321.

4. *Mineral Resources*, 1910, 302

5. Energy Information Administration, *Natural Gas Annual 1985* (Washington,D.C.: Energy Information Administration, 1986), 60.

6. Domestic customers included households or residential customers served and some small commercial establishments. In 1930 the Bureau of Mines began collecting and reporting separately statistics for residential and commercial customers. The terms residential and domestic are used interchangeably in the text.

7. *Mineral Resources*, 1902. Data on the number of fires and the number of customers can be used to obtain an indication of the extent to which NG customers used NG in the different states. First, the number of fires for 1902 is estimated by multiplying the number of fires for 1901, the last year for which the number of fires was reported, by the geometric average of the percentage change in the annual number of fires between 1897, the first year for which the number of fires was reported, and 1901. Second, the estimated number of fires for 1902 is then divided by the number of reported domestic customers for 1902 to obtain an indication of the number of fires per customer. The number of fires per customer were 2.23, 2.14, 1.99, 1.67, 1.43, and 1.39 in West Virginia, New York, Pennsylvania, Kansas, Ohio, and Indiana, which states accounted for 99% of the total value

of NG consumed in the United States in 1902. The number of fire statistics also indicate how dramatically the size of the market was changing between years in different states near the turn of the century. The Kansas market was very small in 1897 with 3,956 fires, and the Indiana market was the largest with 214,750 fires. However, there were only 153,869 fires in Indiana by 1901. By 1901 Pennsylvania with 326,912 fires had twice as many fires as any other state, while Ohio, West Virginia, and Kansas had 149,709, 55,508, and 10,227 fires, respectively.

8. *Mineral Resources*, 1906. In 1906 domestic consumption by state was first reported. It is assumed that the number of fires per customer was the same in 1902 and in 1906 to obtain the estimate of 75Mcf.

9. American Gas Association, *Historical Statistics of the Gas Utility Industry 1966-1975* (Arlington, Va.: American Gas Association, 1977), 148 (hereafter, *Historical Statistics, Gas*).

10. The following texts suggest that NG was used only for lighting and cooking in households. A. R. Tussing and C. C. Barlow, *The Natural Gas Industry: Evolution, Structure and Economics* (Cambridge: Ballinger Publishing Company, 1983). A. M. Leeston, J. A. Crichton, and J. C. Jacobs, *The Dynamic Natural Gas Industry, The Description of an American Industry from the Historical, Technical, Legal, Financial and Economic Standpoints* (Norman, Oklahoma: University of Oklahoma Press, 1963). Photographs and several texts suggest that gas satisfied water heating, space heating, and other needs of households. A. Lief, *Metering for America* (New York: Appleton-Century-Crofts, Inc., 1961), 1-107. L. S. Russel, *Handy Things to Have Around the House* (New York: McGraw-Hill Ryerson Limited, 1979). D. Hale, *Diary of an Industry* (Dallas, Texas: American Gas Journal, 1970). Also see the sources cited in references 24 and 39.

11. H. McCullough, and M. Brignano, *The Vision and Will to Succeed, A Centennial History of the Peoples Natural Gas Company* (Pittsburg, Pennsylvania: The Peoples Natural Gas Company, 1985), 65.

12. To estimate the effect of the price of NG on NG demand is much more difficult for the industrial sector than for the residential sector. Industrial use was greatly affected by activity levels in particular industries and by the price of substitute fuels for which annual data are not available. Even today it is much more difficult to obtain reliable estimates of price effects for the industrial sector. D. R. Bohi, *Analyzing Demand Behavior, A Study of Energy Elasticities* (Baltimore: Johns Hopkins University Press, 1981), 105-113. N. D. Uri, *The Demand for Energy and Conservation in the United States* (London: JAI Press Inc., 1982), 53-91.

13. U.S. Bureau of the Census, *Historical Statistics of the United States, Colonial Times to 1970* (Washington, D.C.: U.S. Government Printing Office, 1975), 105. (hereafter, *Historical Statistics, Census*).

14. Bureau of the Census, *Census of Manufacturing 1914*, States (Washington, D. C.: U. S. Government Printing Office, 1916), Tables 2, 7, and 9.

15. For photographs and newspaper accounts see M. Brignano and H. McCullough, *The Spirit of Progress, The Story of East Ohio Company and the People who Made It* (Cleveland, Ohio: The East Ohio Gas Company), 1-25.

16. Brignano and McCullough, *Spirit of Progress*, 17-18.

17. *Mineral Resources*, 1911, 285.

18. The percent of the state population living in cities and urban areas may have been more important than density of population for the growth of the residential electrical energy

markets as argued in an article by Woolf. Nonetheless, Woolf did not estimate a strong relationship between the percentage of housing units electrified and either urbanization or density of population. A.G. Woolf, "The Residential Adoption of Electricity in Early Twentieth Century America," *Energy Journal*, 8 (1987): 19-30. However, a similar case can not be made for RNG markets. Economies of scale were greater for electric generation plants than for NG distribution plants. Thus, electric plants had more to gain than gas plants from locating in cities. Moreover, the relationship between the development of RNG markets and the development of cities is probably best described as an interdependent rather than a dependent relationship. The availability of gas distribution plants may have contributed as much to the further urbanization of a region as urbanization did to this development.

19. *Mineral Resources*, 1912, 57-59; 1917, 1107-1109.

20. *Mineral Resources*, 1902, 639.

21. *Mineral Resources*, 1913, 1422-1423.

22. *Mineral Resources*, 1906, 821-822.

23. *Mineral Resources*, 1910, 314-315.

24. Rose, "Urban Environments," 503-539. M. H. Rose and J. G. Clark, "Light, Heat, and Power: Energy Choices in Kansas City, Wichita, and Denver, 1900-1935," *Journal of Urban History*, 5 (1979): 360-364.

25. *West versus Kansas Natural Gas Co.*, 221 U.S. 229 (1911).

26. *Mineral Resources*, 1911. A discussion of the situation in Kansas can be found in J. G. Clark, *Energy and the Federal Government* (Chicago: University of Illinois Press, 1987), 21-22, 35-36, 85, 88-91.

27. G. L. Wilson, J. M. Herring and R. B. Eutsler, *Public Utility Industries*, (New York: McGraw-Hill Company, Inc., 1936), 78.

28. For discussions of utility rate-making for electric, gas, water, telephone, and street railway service see W. E. Mosher and F. G. Crawford, *Public Utility Regulation* (New York: Harpers & Brothers Publishers, 1933), 262-316. For electric utility rate setting see W. J. Hausman and J. L. Neufeld, "Time-of-day Pricing in the US Electric Power Industry at the Turn of the Century," *Rand Journal of Economics*, 15 (Spring 1984):116-126.

29. T. J. Tonkin Jr., "The Menifee (Kentucky) Storage Field," in *Proceedings of the Natural Gas Section of the American Gas Association* (New York: American Gas Association Annual Meetings, 1940), 192-193 (hereafter, *American Gas Association, Proceedings*).

30. *Mineral Resources*, 1922, 346.

31. *Pennsylvania versus West Virginia*, 262 U.S. 553 (1923).

32. E. Troxel, "Regulation of Interstate Movements of Natural Gas," *Journal of Land and Public Utility Economics*, XIII (February, 1937): 20-30. Besides the statements of Holmes and others, this article has a very useful summary of judicial opinions for the earlier period.

33. Particular examples of such changes can be found in A. F. Dawson, *Columbia System, A History* (New York: J. J. Little and Ives Company, 1937).

34. Statistics are not generally available on the amount of MG and of NG used in households of the United States in the first decade of this century.

35. This number is the sum of coal gas produced at gas works (35 Bcf) and of oil and water gas (123 Bcf). *Mineral Resources*, 1912, Gas, Coke, Tar, and Ammonia Chapter,

1161-1181.

36. This number is the sum of coal gas, water gas, and oil gas from *Mineral Resources*, 1920, Manufactured Gas and By-products chapter, 439-497.

37. Quoted in J. C. G. Conniff and R. Conniff, *The Energy People, A History of PSE & G*, (Newark, New Jersey: Public Service Electric and Gas Company, 1978), 35.

38. In 1908, more than 100 Bcf of MG was obtained from a process that used both oil and water. This gas was well suited for cooking. Customers who purchased this gas in 1908 used 76.9 Bcf for illuminating purposes and 26.5 Bcf for fuel purposes. In 1912, 68.1 Bcf was sold for illuminating purposes and 54.6 Bcf was sold for fuel purposes. *Mineral Resources*, 1912, Gas, Coke, Tar, and Ammonia Chapter, 1161-1181.

39. This section draws on information from a variety of sources. H. J.Cowan, *Science and Building, Structural and Environmental Design in the Nineteenth and Twentieth Century* (New York: John Wiley & Sons, 1978), 215-296. E. Mayhew, E.& M. Myers Jr., *A Documentary History of American Interiors, From the Colonial Era to 1915* (New York: Charles Scribners & Sons, 1980). B. Nessel, "Uncontrolled Fire to Regulated Heat, A History of Heating," *Trade Winds*, Vol. 5, No. 6 (Minneapolis, Minnesota: Honeywell Regulator Company, 1980).

40. L. Stotz, *History of the Gas Industry*, (New York: Stettiner Bros.,1938).

41. Because data for Kansas reflect special circumstances for several years, these data are not examined with the other states. The five states are designated key states and are treated as a panel to be examined over time.

42. For temperature data see *Historical Statistics, Census*, Series J, 441-447. Temperature data were not available for each state but only for selected sites. The temperatures at these sites do not necessarily represent the ambient temperature experienced by NG customers in the state. For example, the New York temperature does not represent the temperature for NG customers who resided in the southwest portion of the state, but the temperature for New York City customers where MG customers lived. However, changes in the temperature between years are considered to be similar in the two parts of the state. The same consideration is used in assigning changes in temperature at a site with changes in temperature for a state. For Ohio and Pennsylvania temperature from Wooster Experiment Farm, Ohio was used. For West Virginia, the Urbana (U. of Ill.), Illinois temperature and for Oklahoma the Goodwill Agricultural College, Oklahoma Station was used.

43. An additional equation was also estimated to examine the individual state effects. This equation (standard errors are in parenthesis) is:

$$QD = 234Mcf -0.65(PG) -65(DT) -18Mcf(OK)$$
$$(26) \quad\; (0.48) \quad\;\; (24) \quad\;\; (4)$$

$$-29Mcf(PA) -37Mcf(NY) -47Mcf(OH)$$
$$(5) \qquad\quad\; (7) \qquad\quad\;\; (6)$$

$R^2 = 0.83$ 1st order autocorrelation $= -0.082$

The intercept term is an estimate of the level of consumption in West Virginia after correcting for the effect of price and temperature on use. The price effect is much different from the initial estimation. This points up one of the principle and inescapable difficulties for the estimation. There is a lack of independence between state effects and price effects. If we estimate state effects for all states we are possibly capturing in the estimated state effects, effects that are due to price.

44. The price elasticity was calculated as the product of the regression coefficient and the ratio of the average price of natural gas to the average quantity of natural gas consumed per customer for the period.

45. Some of the difference in state effects may have also been due to such factors as differences in the type of appliances in the state. Natural gas was available to a very large number of communities in Ohio, which may have increased the competition between sellers of natural gas appliances resulting in the availability of better, more efficient appliances.

46. The procedure for obtaining the range of values or bounds is described in Herbert, "A Data Analysis," and in J. H. Herbert, and K. Dinh, "Reporting the Uncertainty in Regression Coefficients from Errors-in-Variables," in *Proceedings of the Section on Economic and Business Statistics* (Alexandria, Virginia: American Statistical Association, 1988), 259-264. The properties of the estimator as well as the model upon which the procedure is based is delineated in R. Kalman, "System Identification." The range of values reported almost certainly bound the true coefficient defining the relationship between the aggregate use of NG per domestic customer and the average price of NG for the population of domestic customers in the six states between 1910-1919. A major assumption underlying the model is that the random measurement errors and the left-out explanatory variables are independent of each other and are independent of the true or correctly measured variable. Therefore, if there is a strong linear relationship between a left-out variable and the average price of NG the reliability of the bounds for the price coefficient is questionable. Information about the market suggests that income, coal price, and electricity price influenced average use per customer during the period. Appropriate state level values for these variables are not available. However, examination of available data on national consumer price indices for the price of electricity (beginning in 1913) from the United States Department of Labor, Bureau of Labor Statistics, on state and national coal prices for coal production from the United States Department of the Interior, United States Geological Survey and on national income series from the United States Bureau of Census indicate that there is little relationship between these series and the NG price series.

47. Lengthier discussions and graphical treatments of the relationship between estimated elasticities and gains to buyers and sellers using natural gas markets as a case study can be found in several papers. J. H. Herbert, "The Teaching of Demand Using Actual Data and Applications," *Economics*, 23 (November 1987):3-5. J. H. Herbert, "Combining Economic Ideas and Economic Data in Introductory Economics Courses," *The Northern Virginia Review*, 4 (Spring 1988):19-23. This would be an especially appealing strategy if most of the customers in a region were already gas customers and thus there would be little risk in losing new customers because of the increase in price.

48. Ten percent of all American dwelling units used NG. In Ohio alone, 890,000 customers, 63% of all dwelling units, used NG. The statistics on dwelling units are from United States Department of Commerce, Bureau of the Census, *Census of Housing 1920* (Washington, D.C.: U.S. Government Printing Office, 1924). The statistics on number of customers are from Mineral Resources, 1920. Although one customer, according to the United States Department of Interior, Bureau of Mines definition is not necessarily equivalent to one dwelling unit, according to the United States Department of Commerce, Bureau of Census definition of dwelling unit, they are similar measurements and the ratio

of the number of customers to the number of dwelling units indicates the extent to which domestic natural gas markets had developed.

49. T. K. McCraw, *Prophets of Regulation*, (Cambridge, Massachusetts: The Belknap Press of Harvard University Press, 1984), 1-61.

50. Clark, *Energy and the Federal Government*, 35.

51. D. Schap, *Municipal Ownership in the Electric Utility Industry*, (New York: Praeger Special Studies, 1986), 19-48. E. Troxel, *Economics of Public Utilities*, (New York: Rinehart & Company, 1947), 49-88.

52. *Mineral Resources*, 1920, 291.

53. D.B. Dow, *Effects of Gasoline Removal on the Heating Value of Natural Gas*, Technical paper 423 (Washington, D. C.: Bureau of Mines, U. S. Department of Interior, 1920).

54. *Mineral Resources*, 1920, 290.

55. The cost of NG to the natural-gas gasoline plant was $0.84 per Mcf. Each Mcf of NG yielded .78 gallons of gasoline, which yield had significantly increased from 1916, when it was .50 gallons/Mcf, and each gallon of gasoline was valued at 18.7 cents a gallon at the plant. For additional detail see *Mineral Resources*, 1920, 290.

3

CHANGES IN THE MARKET
DURING THE 1920's

Despite the fact that the NG market had grown at a rate far greater than the growth in GNP in constant dollars between 1910 and 1920, coal and petroleum markets had exhibited even greater growth. At the beginning of the third decade of the twentieth century, prospects for NG markets were considered by some to be extremely uncertain and limited, and analysts at the Bureau of Mines were declaring that "the annual output of natural gas will never be much more than what it was during the period 1916-1920."[1]

Perceptions of the viability of the NG industry would change as the NG market experienced continued growth throughout the 1920's. Yet even though the market would continue to grow, most growth would occur in the West and South, in California, Texas, Oklahoma, and Louisiana, and not in the North during the decade (see Table 3.1).

There was, however, significant market growth in percentage terms in Colorado, Missouri, Wyoming, Montano, and Arkansas, as pipeline systems were extended to and within these states, especially near the end of the decade. For example, pipelines connected the Panhandle field in Texas to Kansas City, Missouri, and Denver, Colorado, during 1928 and 1929. As a consequence, the number of cubic feet delivered to domestic customers in Colorado and to industrial customers in Missouri quadrupled and doubled, respectively, between 1928 and 1929.

The 500-mile pipeline to Denver began in the northwest corner of Texas near the Oklahoma and Kansas borders and served such major Colorado cities as Pueblo and Colorado Springs along the way to Denver. The construction of an interstate pipeline system from Texas to Denver was significant to the NG industry in Texas in that Denver was a hub for the western region made up of Idaho, Wyoming, Montana, Utah, and Nevada, and supplied a lucrative market for Texas NG. The pipeline system was significant for the pipeline company because it was assured control of this growing market for the future by building the pipeline.

Denver contained 28 percent of the population of Colorado and Colorado accounted for more than 36 percent of the population in this region. The pipeline was significant for residential gas customers in Denver because many of them were able to substitute lower cost NG for expensive MG which they had been using extensively in water heaters, ranges, space heaters, and many miscellaneous appliances.[2] By 1930 36 percent of the dwelling units in Colorado were receiving NG.

Table 3.1

Industrial and Domestic Consumption in 1920 and in 1929

	Industrial				Domestic (Residential)					
	Use[a]		Price[b]		Use[a]		Price[b]		Consumers[c]	
	1920	1929	1920	1929	1920	1929	1920	1929	1920	1929
Ohio	40	39	33	42	97	88	38	64	937	1,214
Pennsylvania	95	70	29	35	66	53	37	62	510	634
California	54	299	10	10	12	44	61	87	216	1,239
Texas	40	427	13	10	10	39	44	64	106	503
Oklahoma	107	298	12	10	20	31	31	45	123	225
West Virginia	79	49	16	23	21	25	25	35	131	185
New York	1	2	23	56	18	17	39	67	190	248
Kansas	20	60	15	16	13	16	42	60	116	170
Kentucky	6	8	23	23	10	12	37	46	100	157
Arkansas	14	32	15	12	5	8	38	49	34	62
Missouri	.7	8	23	20	5	7	73	86	76	121
Louisiana	41	193	6	6	5	4	18	54	22	115
Montana	.7	6	2	10	.2	3	48	43	1	15
Wyoming	9	41	5	5	1	3	34	47	2	15
Colorado	.0	12	–	17	.0	3	13	25	–	77
Indiana	2	2	38	53	3	.7	45	65	35	16
Illinois	3	3	11	11	.4	.1	46	84	4	13
Other	.0	11	46	18	.5	18	57	78	12	104

Sources: *Mineral Resources*, 1920, 353, 355; *Minerals Yearbook*, 1929, 327, 329.

[a] Consumption in billion cubic feet (rounded).

[b] In cents per thousand cubic feet(rounded).

[c] In thousands of consumers(rounded).

[d] The other states are Alaska, Iowa, Maryland, North Dakota, Washington.

The many pipeline extensions that were built or planned near the close of the 1920's indicate the level of industrial maturity of the NG industry at the time (see Table 3.2). Several pipeline systems approximately 500 miles in length were completed in 1929 alone, and at least one pipeline system more than 1,000 miles in length was planned.

The 1920's were to be great years for industrial customers served by the NG industry, many of whom were able to obtain NG at a relatively low cost.[3] They were not to be great years for domestic customers. Many domestic customers were only able to obtain NG at a constantly increasing cost since institutional and economic conditions for obtaining a ready supply of NG at least cost favored industrial, not residential customers.

Institutions such as public service commissions frequently perceived the public interest to be best served by allowing utilities to operate with a minimum of regulation.[4] Discounts on NG were occasionally provided to stockholders and managers of utility companies, industrial suppliers of goods and services for the utility, and to large volume industrial customers.[5] At this time, holding companies, over whose activities public utility commissions had very little control, were also near their apex.

Holding companies were businesses that could hold financial interests in a wide assortment of firms such as gas utilities, steel plants, coal companies, and smelting operations.[6] The control of business relationships by holding companies was very likely to serve particular interests. For example, a manufacturing plant controlled by a holding company, which also had control of a NG production or pipeline company, might be able to obtain NG at a large discount. This would be especially likely to occur if this reduction in the cost of operation helped it to cut the price of its product and to drive out of business a competing company.

While the average real cost of NG in the United States between 1920 and 1929 fell from $.17 Mcf to $.12 Mcf for industrial customers, it rose from $.38 Mcf to $.62 Mcf for residential customers. Accordingly, consumption per domestic customer fell nationally from 109 Mcf to 70 Mcf per year, and aggregate domestic consumption increased by only 26 percent. Aggregate industrial consumption increased by more than 200 percent. Although the number of domestic customers in the five key states of West Virginia, Ohio, Pennsylvania, New York, and Oklahoma, examined in detail in the previous chapter, grew by 33 percent between 1920 and 1929, sales volumes declined by 3.3 percent. Much of this decline may have been a response to the increase in the cost of NG.

Now several explanations were offered during the decade for the high cost of NG to domestic customers. The depletion of NG wells in such states as Ohio was frequently offered as a reason for the large price increases at the beginning of the decade in that particular state. In fact, some NG companies such as the East Ohio Gas Company were beginning well-publicized conservation education programs as early as 1920.[7] However, the above average cost of NG gas to domestic customers in several producing states (see Table 3.1) was clearly not a consequence of the depletion of wells. The increase in the number of residential customers in such states as Pennsylvania and New York was also not consistent with an overall depletion of existing wells. Finally, the percentage increase in

Table 3.2

Pipelines Completed, Under Construction, or Planned during 1929

Status and Origin[a]		Length and Destination[b]
C	Monroe-Richland, LA.	140 Shreveport, LA.
C	Monroe-Richland, LA.[c]	450 Atlanta, GA.
C	Monroe-Richland, LA.	450 St. Louis, MO.
C	Lea, NM.	200 El Paso, TX.
C	Bruni-Mirando, TX.	120 San Antonio, TX.
C	Panhandle, TX.	350 Dallas-Fort Worth, TX.
C	Panhandle, TX.	450 Kansas City, MO.
C	Jennings, TX.	165 Monterrey, TX.
C	Kettleman Hills, CA.	300 San Francisco, CA.
C	Kettleman Hills, CA.	100 San Joaquin Valley, CA.
C	Baxter Basin WY.[d]	250 Salt Lake City, UT.
C	Clarksville, AK.	100 Little Rock, AK.
U	Kettleman Hills, CA.	400 San Francisco-Sacramento, CA.
U	Glendive, MT.	175 Willston, ND.
U	Clarksville, TX.	400 Nashville, TN.
U	San Juan, NM.	200 Albuquerque-Santa Fe, NM.
U	Medicine Lodge, KS.	360 Grand Island, NB.
U	Maljamar, NM.	140 Clovis, NM.
U	Oklahoma City, OK.	60 Perry, OK.
U	Atchinson, KS.(extension)	60 Falls City, NB.
U	Lyons, KS.(extension)	120 Superior, NB.
P	Kettleman Hills, CA.	800 Oregon-Washington
P	Ogden UT.(extension)	125 Pocatello, ID.
P	Panhandle, TX.	950 Chicago, IL.
P	Panhandle, TX.	1250 Indianapolis, IN.
P	Los Angeles Basin, CA.	100 San Diego, CA.
P	Glendive, MT.	250 Bismarck, ND.
P	Glendive, MT.	600 Minneapolis-St. Paul, MN.
P	Big Horn Basin, WY.	260 Anaconda-Butte, MT.
P	Hugoton, KS.	450 Omaha, NB.-Council Bluffs, IA.
P	Kentucky	300 Detroit, MI.
P	West Virginia	400 Richmond, VA.-Washington, D.C.
P	West Virginia	350 Philadelphia, PA.
P	Montana	700 Coeur d'Alene, ID.-Spokane-Seattle, WA.

Source: *Minerals Yearbook*, 1929, 340.

[a] C, U, and P indicate completed, under construction, or planned.

[b] Length in miles.

[c] Branches to Jackson, Selma, Macon, Mobile, and other cities were either under way or contemplated in 1929.

[d] Also from the Hiawatha Basin in Colorado.

price was very similar in states such as Ohio where deliverability problems were reported and states such as Oklahoma where supplies were generally more than adequate.

Some of the increase in price to domestic customers during the twenties may have been due to depletion effects, but if the own-price elasticity for natural gas demand per customer was inelastic during the 1920's, as it appears to have been during the 1910's, a major proportion of the increase may have been due to the price setting behavior of NG businesses. To repeat: NG companies could maximize profits between years by raising the price to their existing domestic customers. They could also subsidize industrial customers with economic rents gained from domestic customers.

The large price increases during the 1920's may have not only increased the profitability of gas companies but, depending on the exact magnitude of the own-price elasticity, may have significantly reduced the economic welfare of domestic consumers. As the price of NG rose, many NG consumers could very well have been allocating an increasing proportion of their income to the purchase of NG services since their real income was relatively constant during the period. The average (geometric) growth in real income per capita between 1919 and 1929 was only 1.9 percent. Moreover, the distribution of income was becoming more unequal as a larger proportion of any growth in income was being allocated to upper income households.[9] Thus many of the workers in Ohio who used NG in their homes were most likely experiencing a reduction in their purchasing power because of large NG price increases, especially during the first half of the 1920's. Any reduction in purchasing power was especially significant because it occured at a time when an increasing number of new consumer products were being marketed, and, thus, the increase in the price of NG effectively cutoff these consumers from some of these markets. The steep rise in the price of NG to domestic customers and the decline in use per customer were distinguishing features of RNG markets during the 1920's.

GAS USE PER CUSTOMER, PRICE, AND WEATHER

In order to examine more generally the effect of price on NG demand per customer(QD), an equation was estimated by regression techniques similar to the equation estimated in the previous chapter. However, instead of separate state effects for particular states, effects are specified for pairs of states, primarily because of perceived similarities in the availability of NG within these states during the time period. These pairs of states are West Virginia and Oklahoma (WV/OK), and Ohio and New York (OH/NY). Oklahoma, where the pipeline infrastructure had improved significantly since 1910, was combined with West Virginia to represent states in which NG was most generally available. Oklahoma and West Virginia were the major exporters of NG during the 1920's, and large supplies of NG were available to domestic customers within these states.

On the other hand, Ohio and New York were major importers of NG. In these states deliverability problems in the wintertime were reported. Pennsylvania was neither a major importer nor a major exporter relative to the other states, and, thus, this state was average in terms of the availability of NG for the time period.

Table 3.3

Average Price and Amount Sold of Natural Gas during the 1920's

	West Virginia	Oklahoma	Pennsylvania	Ohio	New York
QD^a	146.5	122.8	100.2	82.3	76.7
PG^b	35.2	53.5	61.8	62.3	66.4

Sources: Quantity and Price, *Minerals Resources*, 1920-1924, various pages, *Minerals Yearbook*, 1925-1929, various pages; Deflator, Consumer Price Index - All Items, *Historical Statistics, Census*, 211.

  ^{a} Average quantity consumed per domestic customer in thousand cubic feet.

  ^{b} Average price per thousand cubic feet.

Some idea of the effect of price on use per customer can be gleaned directly from data on price and use per customer (see Table 3.3). A state with a higher average price always had a lower average use per customer. Some idea of the effect of availability on the level of use per customer between states can also be gleaned directly from the data in the case of Pennsylvania and Ohio. The average change in temperature between years and the average price of NG was essentially the same in these two states. In addition, examination of aggregate data on oil and coal prices revealed that these prices were similar in these two states. Thus, the difference in the average values for QD of 18 Mcf annually, which was approximately the amount of NG used by a large range or a small water heater during the time period, supplies an indication of the effect of availability on the amount of NG used by the average customer. Since available but incomplete evidence on income per capita for Ohio and Pennsylvania indicates that income per capita was less in Ohio during the time period, it is also necessary to assume that this difference between states had little or no effect on the level of NG consumption per customer.

Although the absolute magnitude of the price effect (see Table 3.4) was lower when compared to the same effect in Table 2.8, the own-price elasticity for the 1920's was higher and was equal to -0.53 with bounds of -0.53 and -0.64.[9] Thus, the elasticity falls within the bounds for the elasticity estimated in the previous chapter. Moreover, the magnitude of the elasticity coupled with the large increase in price and the modest increase in income, which occurred during the time period, strongly suggests that many RNG customers were experiencing reductions in their purchasing power during the time period.

Table 3.4

Estimated Effects

	Price	Temperature	State Effect WV/OK	OH/NY
Effect	-1.10	63	15	-18
Standard Error	.08	21	3	3

$R^2 = 0.96$, 1st order serial correlation $= 0.27$

Even though price influenced the size of the RNG market through its influence on use per customer, other factors, in addition to price, had an influence on the size of the market as measured by the number of customers. One important factor was the control of gas markets by MG interests.

MANUFACTURED GAS INTERESTS

MG interests held court in such areas north and east of the Northern NG region as Michigan and the metropolitan areas of Philadelphia, Pennsylvania and New York. Gas utilities in these places had large amounts of capital invested in equipment for producing MG. Moreover, some of the companies operating in these areas, such as United Gas Improvement (UGI) Company of Philadelphia, had been closely associated with major innovations of the MG industry such as the Welsbach lamp and the Lowe process for manufacturing gas. In fact, UGI had been incorporated in 1882 in order to introduce the Lowe process. UGI had also brought the Welsbach lamp to the United States in 1887 and continued to be involved in the marketing of this lamp in the 20th century.[10] The Lowe process was much more efficient than many other processes and eventually became very popular in the United States. The Welsbach lamp had also slowed down the encroachment of electrical energy into the gas lighting market because of its much improved lighting characteristics over earlier gas lamps.

If NG was delivered into MG regions at cost during the 1920's, significant reductions in revenues received by MG utilities would have ensued because NG was not only less costly than MG but also contained about twice as much useful heat per cubic foot. Naturally, the utilities were not willing to scrap their MG equipment before it was fully depreciated because this would not yield them any possible advantage. There were few competitors threatening to sell NG into their regions and drive them out of business. In fact, many of the NG companies, the MG companies and the electric utility plants that could serve the utilities service area were frequently part of the same holding company.[11]

MG utilities were assured a relatively constant revenue stream from one year to the next. They were also able to use their strong economic base to obtain special privileges and to manipulate financial markets. They were also familiar with city government bureaucracies since they had to work with these bureaucracies in order to continue to receive the rights that were necessary to effectively operate utilities in metropolitan areas. It is not surprising that the first state utility commissions were founded in the MG states of New York and Wisconsin.

Charles Evan Hughes in an investigation of the Consolidated Gas Company in New York City near the turn of the century found "gross abuse of legal privilege in overcapitalization and in manipulation of securities for the purpose of unifying control and elimination of possible competition. . . . the investment of millions in securities earning no interest and intrinsically worthless solely for the purpose of securing a monopoly of control."[12] As a consequence of publicity from this investigation, Hughes was elected New York governor in 1907. He then established the first state utility commission.

THE REGULATORY CONTROL OF STATE UTILITY COMMISSIONS

The state of Wisconsin under the control of the populist governor La Follette also founded a state utility commission in 1907. By 1920 thirty five states had state utility commissions with administrative and judicial roles. Between 1910 and 1920 two of the tasks that kept the staffs of these state commissions busy were the designing of standards for the valuation of utility properties and the identification of the economic and equity objectives of a variety of rate schedules. Both of these factors needed to be well understood before standard procedures for evaluating the reasonableness of rates to domestic customers could be established.[13]

By the middle of the 1920's, it was clear to many students of the utility industry, that state commissions were unable to determine whether domestic customers were receiving gas at rates determined by the disinterested or objective application of engineering and economic science. Instead, rates were governed by the long-term strategies of holding companies that had managed during the first several decades of the twentieth century to acquire the engineering, accounting, and legal talent to identify and solve many of the problems involved in supplying electric and gas service to a large proportion of the American public. State utility commissions, however, had not hired the analytic talent to identify whether these solutions might be least costly from the point of view of the domestic consumer. In fact, the setting of rates was considered to be the prerogative of management and only in rare instances did regulators become directly involved in the evaluation of rates especially with respect to NG.[14]

The uncritical way in which state utility commissions viewed the utility companies is understandable. These companies were making utility services of many kinds available at a phenomenal rate to an ever larger proportion of the American public. The availability of electric appliances, telephone service, street cars, and clean, cheap, gas heat were symbols of the good life in America. Utility services, along with motor cars, represented the significance of American society as a culture to the rest of the world.

The great growth in utility services appeared to reflect the relative ease at which holding companies acting for individual utility companies could appropriate and organize financial and natural resources over a wide area to serve the public. In particular, the availability of gas and other utility services enabled an ever larger proportion of the American population to live decent and interesting lives in metropolitan areas and to further specialize their daily economic activities. With a variety of utility services available to them, city dwellers were better able to allocate an increasing proportion of their day to industrial and commercial activity and a decreasing proportion of their day to domestic concerns.

URBANIZATION

A factor that affected the growth of the domestic NG markets and was in turn affected by this growth during the 1920's was the increasing trend in the proportion of the population living in urban places. This trend would continue to influence the growth of RNG markets in later years. A larger proportion of the population was living in urban places rather than in rural places for the first time during the 1920's and nearly 80 percent of the new dwelling units were started in urban locations (see Table 3.5).

Table 3.5

Percentage of New Dwelling Units Started in Urban and Rural Places

	1890-1899	1900-1909	1910-1919	1920-1929	1930-1939
Urban	56.5%	65.4%	73.0%	79.8%	62.4%
Rural	43.5%	34.6%	27.0%	20.2%	31.6%

Source: L. Grebler, D. M. Blank, and L. Winnick, *Capital Formation in Residential Real Estate, Trends and Prospects* (Princeton: Princeton University Press, 1956), 99.

Urban development could have resulted in great growth in RNG markets because the additional capital cost of serving an additional domestic customer was relatively modest. Many of these urban houses were row houses or apartment houses and the cost of putting in a gas line or a series of gas lines for a block of new row houses was of course less than for a detached home. However, the price of NG was becoming less competitively priced in many cities during the time period. In fact, among the sources of energy available to households, only the cost of NG was rising.

THE COST OF NATURAL GAS RELATIVE TO THE COST OF OTHER SOURCES OF ENERGY

Coal, coke, and wood were generally less expensive than NG during the 1920's and coal clearly dominated residential energy markets (see Table 3.6).

Table 3.6

Energy Use in Domestic Markets in 1929 (million short ton equivalents)

	Amount	Percentage of Total
Bituminous Coal	70	43
Anthracite Coal	55	34
Natural Gas	16	10
Coke and Briquets	12	7
Fuel & Furnace Oils	10	6
Total	163	100

Source: National Industrial Conference Board, Inc., *The Competitive Position of Coal in the United States* (New York: National Industrial Conference Board, 1931), 131.

Amongst fuels only MG was clearly more expensive than NG, and even the cost of MG was declining.

The average price of electrical energy to domestic customers was also falling consistently throughout the 1920's, as it had been falling since the turn of the century. Even though the cost of electrical energy was greater than the cost of NG, the expectation of many consumers was that the cost of electricity would continue to decline. Thus, the cost of operating an electrical appliance was expected to decline as well.

The reduction in the cost of electrical energy to domestic consumers was consistent with the block schedule rates that were relatively common for electric service but uncommon for NG service. With block schedules the cost of energy falls as the level of consumption increases. These schedules were much more common for electric service than for NG service, because economies of scale associated with NG distribution were much less than they were for electric generation.

By constantly increasing the price of NG to domestic customers, and by not extending pipelines much into MG regions, the NG industry was unable to take advantage of the growing demand for a convenient, clean, and flexible source of energy in urban areas during an era in which households had access to an ever-expanding number of improved energy-using appliances for the household. Consequently, even though more NG was used in households than fuel oil and coke in 1929 (see Table 3.6), the percentage gain in the use of such fuels as fuel oil and coke after 1924 was greater than the percentage gain in the use of NG (see Table 3.7), even though the sales of NG to residential customers began to rise dramatically at the close of the 1920's.

Table 3.7

Trends in Domestic (Residential) Fuel Consumption, 1924-1929

	1924	1925	1926	1927	1928	1929
Anthracite[a]	57.4	47.7	57.8	54.8	57.4	54.6
Coke & Briquets[b]	6.7	8.9	9.2	8.6	10.3	12.1
Oil for heating homes[c]	5.0	9.0	9.1	11.7	14.3	17.6
Natural Gas[d]	285	272	289	296	321	360

Source: Conference Board, *Competitive Position of Coal*, 137.

[a] Net consumption of anthracite in the United States less quantities used in (1) manufacturing, which are estimated in part, (2) locomotives, and (3) mining for steam and heat. The figure in million tons includes some industrial consumption for fuel at electric plants and other uses.

[b] Fuel Briquets (domestic and imported), by-product and beehive coke, gas-house coke, and petroleum coke in million tons.

[c] In million barrels.

[d] In billion cubic feet.

NATURAL GAS APPLIANCES IN THE HOUSEHOLD

Along with increased urbanization, increased electrification of the household[15] and increased social and economic opportunities for women outside the household, the quality and number of appliances for the home increased. These changes affected the economic and social organization of the household.[16] Upper middle income women or their cleaning help were being emancipated from the dirtiest and heaviest of household chores.

Urban male professionals were commonly depicted in advertisements for central space heating units and other appliances as smart, urbane investors who purchased NG equipment to provide an increasingly efficient and improved household for their family. Yet the NG industry captured a relatively small proportion of the growing market for improved space heating units and household appliances, except for ranges, during the first half of the 1920's. Ranges sold especially well during the latter half of the 1920's because a new style gas range appeared. It was fully enameled and came in a variety of shades.[17] Ranges with exposed burners and black metal high ovens were replaced by streamlined colorful cabinet enclosures. Insulated, more efficient ovens soon followed. Time and temperature control settings and an improved automatic lighter were also added.

More water heating units also appeared in stores during the 1920's, yet relatively few of these gas water heaters were sold at first. In fact, Jacob B. Jones, a utility executive from southern New Jersey, was still trying to sell the gas utility industry on the marketability of gas water heaters at their annual meeting in 1923[18] with the following anecdote: "The prisoners of the Atlantic City jail at May's Landing entered into a strike because they could not have their hot water bath regularly. Nobody ever heard tell of the prisoners in the Cumberland jail at Bridgeton going on a strike - we have a big automatic gas water heater in there and our jail is full all the time."

Even in 1926 when approximately 50,000 automatic water heaters were installed, practically all of these heaters used oil as a source of energy.[19]

EFFORTS TO MARKET NATURAL GAS EQUIPMENT FOR THE HOUSEHOLD

Special discounts and credit terms to buyers and advertisements in magazines and newspapers were being used at the end of the decade to increase sales of space heating and water heating units. Hot water was being sold as a relief from worries about hygiene and as a release from the inconvenience of not having hot water readily at hand when it was most needed. Utilities used trained personnel operating out of Home Service units within the utility to publicize the speed at which hot water and other energy services were available from gas appliances. These personnel helped existing customers and prospective customers in the use and maintenance of gas appliances. Well-decorated windows with appliances appeared in downtown shops, and some displays were paraded in the streets. Demonstrations were also given, attended by hundreds of people.[20]

Plumbers were also contacted by utilities and told about the advantages of gas appliances. Salesmen were also offered special cash incentives to sell gas appliances.

The selling job was made difficult by the technical kinks and the lack of standardization in such appliances as automatic water heaters. The lack of standardization prompted the National Bureau of Standards[21] to recommend an 88 percent drop in the types of water heating units.

Water heating units were also quite expensive. For example, the average cost of an automatic NG water heater installed in 1929 was $144.43, with a range of $60.00 to $400.00.[22] This was approximately the cost of a much-improved automatic water heater in 1975. Accordingly, many customers still managed by heating water with attachments built onto furnaces and stoves or heating the water on the stove itself in a large pot.

By far the best way for utilities to increase revenues from domestic customers was to encourage them to become NG central space-heating customers (see Table 3.8).

Table 3.8

Natural Gas Central Space Heating Units Sold in 1928 and 1929

	Number of Units Sold		Estimated Consumption[f]	
	1928	1929	1928	1929
California[a]	31,690	32,690	144	115
Ohio[b]	18,615	23,539	247	250
Missouri[c]	2,447	4,521	260	212
Maryland[d]	30	133	287	252
Kansas[e]	12	80	290	152

Source: American Gas Association, *American Gas Association, Proceedings 1929* (New York: American Gas Association, 1929) 622-623.

[a] The California utility was Los Angeles Gas and Electric Co.

[b] The Ohio companies were Union Gas and Electric Co., Union Light and Power Company, Columbia Gas and Fuel Company, Northwestern Ohio Natural Gas Company, and Federal Gas and Fuel Company.

[c] Missouri companies were Kansas City Gas Company and Joplin Gas Company.

[d] The Maryland company was Cumberland and Allegheny Gas Company.

[e] The Kansas company was Kansas Electric and Power Company.

[f] In thousands of cubic feet.

The average amount of NG used in central space-heating units varied greatly by state. Yet, in all parts of the country the amount of NG used to obtain space heat from central space-heating units was much greater than in its use in any other household equipment (see Table 3.9). However, few NG central space-heating units were sold relative to the number of NG customers in a state until the end of the decade.

Table 3.9

Estimated Use of Natural Gas in Different Appliances in 1929

	Swanson	Committee
Irons	.7	
Gas Refrigerators	10.2	18.0
Incinerators		18.0
Cabinet Range		18.0
Combined Range and Cooker		9.0
Continuous Water Heater		
30 gallons	19.8	
50 gallons	33.0	
Storage Water Heater	23.1	24.0
Automatic Water Heater	23.1	18.0
Tank Heater		7.5

Sources: Swanson, J.K. Swanson, "Domestic Use of Gas Other than Cooking," in *American Gas Association, Proceedings 1929*, 452; Committee, *American Gas Association, Proceedings 1929*, 491.

Note: The estimates in thousand cubic feet presented in the American Gas Association Proceedings were for manufactured gas appliances; therefore, the estimates were reduced by 45% to reflect the greater efficiency of a natural gas appliance. Natural gas per cubic foot contains almost twice as many Btus as manufactured gas.

In order to guarantee the quality of NG appliances the AGA Testing Lab was established in 1925 in Cleveland, Ohio. Personnel at the lab were not only interested in improving the quality of service but they were also interested in developing innovative services. In 1928 the AGA installed fifteen systems to produce cold air for summer air conditioning on an experimental basis.[23] Air-conditioning units, however, would not be marketed until 1935.

Even without proper testing some utilities were very successful in selling relatively new appliances as a consequence of special marketing programs at the close of the 1920's. For example, Brooklyn Union Gas Company of Brooklyn, New York sold approximately 4,000 MG refrigerators in 1928.[24]

INDUSTRIAL NATURAL GAS MARKETS RELATIVE TO RESIDENTIAL NATURAL GAS MARKETS

The direct contribution of NG to the improvement in the economic welfare of domestic consumers during most of the 1920's was relatively small. In fact, until 1927 MG use in households was increasing by more than NG use, and in 1927 both fuels were capturing about equal shares of the residential gas market (see Table 3.10).[25]

Table 3.10

Domestic and Industrial Consumption (in billion cubic feet)

	Domestic			Industrial			
	NG[a]	MG[b]	All	Field	Carbon Black	Electrical Energy	Ratio[c]
1920	286	214	512	202	41	25	2.2
1921	248	209	414	182	51	24	2.9
1922	255	220	508	198	54	27	2.7
1923	277	234	730	343	109	31	3.8
1924	285	239	856	393	157	48	4.7
1925	272	242	916	424	140	46	4.6
1926	289	265	1024	478	131	53	4.6
1927	296	276	1149	549	144	63	5.1
1928	321	275	1247	574	175	77	4.7
1929	360	276	1557	705	262	113	5.1

Sources: Manufactured Gas, Gould, *Output and Productivity in Electric and Gas Utilities*, 1946, 84; All Other Fuels, *Mineral Resources*, 1920-1924, various pages, *Minerals Yearbook*, 1925-1929, various pages.

[a] NG is natural gas.

[b] MG is manufactured gas.

[c] Ratio of the domestic price to the industrial price.

In 1920 residential use of NG was 42 percent and 698 percent greater than the use of NG in field operations and in carbon black manufacture, respectively. By 1929 domestic use was 49 percent less than its use in field operations and only 38 percent greater than its use in carbon black manufacture. Although the relative amount of NG used in the domestic sector in 1920 was cut in half by 1929, the indirect role of NG in electricity, gasoline, and tire production contributed significantly to the economic welfare of consumers.

The growth in the electrical energy industry during the 1920's was another one of the greatest economic growth stories of the twentieth century. The development of the electrical energy industry paralleled the growth in the number of electric appliances for the household, which increased the number of services available within the household. The rate of growth in NG use to produce electrical energy (see Table 3.10) during the 1920's was greater than the rate of growth in the use of any other fuel[26] for electric generation.

The NG industry was also a major contributor to the growing demand for transportation services by gasoline powered vehicles, as the number of vehicle miles travelled increased from 55,000,000 miles in 1921 to 198,000,000 miles in 1929.[27] NG was processed to obtain natural-gas gasoline and NG was also used as a source of energy to withdraw petroleum from wells and to process petroleum at refineries into gasoline.

NG used in natural-gas gasoline production and in oil production was designated as part of field consumption (see Table 3.10). The growth in field consumption explains about half the growth in industrial consumption of NG during the 1920's.

NG was also increasingly used for carbon black manufacture and the growth in NG use in carbon black manufacture explains about 20 percent of the growth in industrial consumption of NG during the 1920's. The availability of carbon black at a low cost to tire manufacturers continued to contribute to the decline in price and improvements in quality of tires. It was found that the life of a tire, in terms of miles travelled, was extended by 20 percent when carbon black was substituted for zinc oxide in tire manufacture during the First World War. This change was also responsible for the black tread tire that has been popular ever since.[28]

The growth in the use of NG in field, carbon black, and electricity operations between 1921 and 1929 was greatly influenced by price. For example, the cost of NG in field and carbon black activities was less than three cents per Mcf in the latter part of the 1920's in Louisiania and Texas. Consequently, the relative amount of NG used in field, electric utility, and carbon black manufacture declined in Ohio and West Virginia and increased in Texas, California, Oklahoma, and Louisiana, where the cost of natural gas was much lower than in the northern states (see Table 3.11).

During the 1920's the huge Monroe field in Louisiana, which comprised approximately 360 square miles of proven NG producing property, and the Panhandle field in Texas were developed to the point where they could readily serve northern markets. Yet the development of nearby industrial markets in the south central United States and the lack of adequate NG underground storage facilities in the North impeded the construction of efficient and flexible NG distribution systems to many northern states. At the close of the 1920's, there were only eight underground storage reservoirs in the United States, and all but one of the reservoirs put in place during the 1920's were in the South and West.

During the 1920's the NG industry appeared to be spending much of its resources on building ever-larger corporate structures by combining several smaller NG businesses into larger entities and by increasing sales to the industrial sector. However, much of this new industrial business would subsequently be lost at the start of the Great Depression as a consequence of the decline in industrial activity.[29] Yet the NG industry would experience renewed growth in RNG markets near the close of the 1920's, as pipelines were finally extended into regions where large amounts of MG were used and as new NG appliances were purchased by an increasing number of consumers.

Table 3.11

Industrial Consumption of Natural Gas by State (in Bcf)

	Field Use		Electric Utility		Carbon Black	
	1921	1929	1921	1929	1921	1929
Oklahoma	56.3	248.0	4.5	6.7		
California	46.4	216.3	3.2	26.2		
Texas	32.7	128.5	1.5	44.0		176.4
Louisiana	10.7	25.0	1.2	18.6	32.1	85.2
West Virginia	10.7	18.8	3.2	.9	15.5	
Kansas	2.7	12.0	2.2	12.6		
Ohio	3.6	5.1	3.9	4.3		
Total	163.1	653.7	18.8	113.3	47.6	261.6

Sources: *Mineral Resources*, 1922, 333; *Minerals Yearbook*, 1929, 357.

APPLIANCE GROWTH AT THE CLOSE OF THE 1920's

The increased number of water heating and central space heating units sold at the close of the 1920's was distinctive. The rate of growth in the purchase of electric appliances had been phenomenal during the first half of the 1920's yet had subsided near the close of the 1920's. This change has suggested to some that consumer satiation may have been reached. This change, along with a decline in the rate of growth of automobile registrations and urban residential construction, has been identified as one of the possible causes of the Great Depression.[30] While consumer willingness to continue to pay for the novelty of electric grills, shaving mugs, and other electric appliances may have abated near the close of the 1920's, the demand for NG water heaters and NG central space-heating units at the close of the 1920's was clearly growing.

As an increasing proportion of the households in the United States obtained NG service in the late 1920's, the general public began to realize that the availability of NG could markedly improve their quality of life. Reports in the newspaper were continually alerting consumers to the opportunities obtainable from receiving NG service. Some of these reports also informed their readership that only through political activism could these opportunities be realized. Events that occurred in Atlanta in 1929, prior to the receipt of NG service there at the beginning of 1930, illustrates these points well. In particular, it exemplifies the value of the substitution of NG service for MG service; a process that was to occur with increasing frequency during the 1930's.

MARKET GROWTH IN GEORGIA AT THE CLOSE OF THE 1920's

In 1929 Southern Natural Gas Company was planning to build a pipeline to Atlanta and to obtain a franchise to operate a distribution company within the city. During February 1929, Southern Natural Gas Company ran a newspaper ad in which it offered to sell NG at a lower price than MG on a volumetric basis.[31] Since NG on a volumetric basis contained nearly twice as much useful energy as MG, this offer would provide immediate and significant savings in energy expenditures to the many households who were using MG and who could switch to NG gas upon the receipt by Southern Natural Gas of a franchise.[32]

The operators of Atlanta Gas Light Company manufactured gas from coal and other hydrocarbons in 1929, and they wanted to be the distribution company for NG as well. After meetings and correspondence between representatives of Atlanta Gas Light and a special committee of the city council, the public service commission and representatives of towns and cities near Atlanta, a contract was signed in which Southern Natural Gas would deliver NG to Atlanta Gas Light Company at the city gate. Thus, Atlanta Gas Light Company would be both the MG company and the NG company in Atlanta.

On November 25, 1929 Atlanta Gas Light Company filed rates for NG with the public service commission which made questionable any savings to consumers from NG service. The public outcry was great. As a consequence the public service commission issued an order for rates that would result in gas consumers receiving savings from converting to NG that would be similar to the level of savings implied by the Southern Natural Gas advertisements. Thus residential customers received NG at a low price only because there initially was competition for the domestic NG market in Atlanta.

The arrival of NG meant actual savings to consumers at the beginning of the Great Depression. It also meant relief from the environmental damage caused by a dependency of industry and households on coal for their energy needs. An editorial in the January 21, 1930 issue of the Atlanta Journal summed up the situation with the following words, "There arrived in Atlanta today a . . . servant of the household . . . . It means cleaner air and skies, quickened and multiplied development of industrial resources, the opening of fresh fields for capital and labor, the harnessing of potent sinews for production and prosperity."[33]

NOTES

1. *Mineral Resources 1922*, 335-368
2. Rose, "Urban Environments," 503-539.
3. Along with the growing importance of the NG industry in the industrial sector, the NG industry was also growing in importance in energy markets. The growth in the value of NG at the mine mouth, referred to as the well-head by the oil and gas industry, was greater than the growth in the value of petroleum and coal, although the total value of NG was still much less than the total value of petroleum and coal. The growth in value of NG production corresponded closely to the growth in one of the major growth sectors among major energy industries in the 1920's, NG-gasoline. In 1920 the value of NG and NG-gasoline was $75,000,000 and $72,000,000, respectively, at the site of production. By 1929, the production of each of these two items had grown by $170,000,000. For data on the value of fuels at the minemouth and for a general discussion of NG, oil, and coal production during the time period, see H. Barger and S.H. Schurr, *The Mining Industries, 1899-1939, A Study of Output, Employment, and Productivity* (New York: American Book-Stratford Press Inc., 1944).
4. Troxel, *Economics of Public Utilities*, 49-88.
5. Nichols, *Public Utility Service*, 902-1020.
6. A. A. Berle and G. C. Means, *The Modern Corporation and Private Property* (New York: Harcourt, Brace & World, Inc., 1968).
7. Brignano and McCullough, *The Spirit of Progress*, 42-44.
8. Analyses relevant to the discussion of income as well as relevant income series for the time period can be found in several texts. P. L. Lee and P. Passell, *A New Economic View of Economic History* (New York: W.W. Norton & Company, 1979), 338-341. J. Hughes, *American Economic History*, (Glenview, Illinois: Scott, Foresman and Company, 1987), 428-433.
9. The results of estimating an equation with dummy variables for all states is reported below (standard errors are in parenthesis):

$$QD = 244Mcf - 1.01(PG) - 63(DT) - 6Mcf(OK) - 19Mcf(PA) - 36Mcf(OH) - 38Mcf(NY),$$
$$\quad (24) \quad\quad (.08) \quad\quad (21) \quad (3.1) \quad\quad (3.5) \quad\quad\quad (3.5) \quad\quad\quad (3.8)$$

$R^2 = 0.96$, 1st order autocorrelation $= 0.22$,

and where,

OK = dummy variable for Oklahoma,
PA = dummy variable for Pennsylvania,
OH = dummy variable for Ohio,
NY = dummy variable for New York,

all other variables have been previously defined.

For this application it is assumed that the amount of NG able to be distributed to domestic customers was not declining systematically between years. If this assumption is false and an increasing number of customers were being curtailed between years, then a decrease in use per customer could be interpreted as a decline in deliverability between years but could not be interpreted as a response to price. In this case a demand relationship could not be estimated from the data. From a reading of the literature, it appears that industrial customers were much more likely to be curtailed than domestic customers and that state utility commissions had a role in ensuring deliveries to residential customers if at all possible. Thus, the increase in price is most probably a consequence of a shifting supply curve due to the increased expense involved in delivering sufficient amounts of NG to markets to satisfy space heating and other energy needs of consumers. With a shifting supply curve, a demand relationship is readily identifiable.

10. N. B. Wainwright, *History of Philadelphia Electric Company 1881-1961* (Philadelphia: Philadelphia Electric Company, 1961), 12.

11. Federal Trade Commission, *Utility Corporations, Report No. 84-A, Final Report of the Federal Trade Commission to the Senate of the United States on Economic, Corporate, Operating, and Financial Phases of the Natural-Gas-Producing, Pipe-line, and Utility Industries with Conclusions and Recommendations* (Washington, D. C.: United States Government Printing Office, 1936).

12. W. E. Mosher, and F. G. Crawford, *Public Utility Regulation* (New York: Harper Brothers Publishers, 1933), 21.

13. E. I. Hellebrandt, "The Development of Commission Regulation of Public Utilities in Ohio," *The Journal of Land and Public Utility Economics* (1933):395-409. This article was continued in the 1934 issue of *The Journal of Land and Public Utility Economics* on pages 78-94. J. Bauer, *Effective Regulation of Public Utilities* (New York: The Macmillan Company, 1925). M. G. Glaeser, *Outline of Public Utility Economics* (New York: The Macmillan Company, 1927).

14. Mosher and Crawford, *Utility Regulation*, 245-246. H. G. Burke, *The Public Service Commission of Maryland* (Baltimore: The Johns Hopkins Press, 1932), 159.

15. G. L. Wilson, J. M. Herring, and R. B. Eutsler, *Public Utility Industries* (New York: McGraw-Hill Book Company, 1936), 92-95. L. Hannah, *Electricity Before Nationalization* (Baltimore: Johns Hopkins University Press, 1979), 186-212. Wainwright, *History*, 186-219.

16. Rose, "Urban Environments," 503-540. Rose and Clark, "Light, Heat, and Power," 340-364. R. S. Cowan, "The Industrial Revolution in the Home: Household Technology and Social Change in the 20th Century," *Technology and Culture* (1976):1-23.

17. Stotz, *History of the Gas Industry*, 169-170.

18. J. B. Jones, "Selling the Gas Man Hot Water," *Gas-Age Record*, (April 14, 1923):465-466.

19. *Heating and Ventilating*, July 1933, The Industrial Press Publishers, 148 Lafayette Street, New York, New York, 17-18.

20. Many examples of the public relations and advertising practices of the gas utilities industry at the close of the 1920's can be found in *American Gas Association, Proceedings 1927-1929*. A. Lief, *Metering for America*, (New York: Appleton-Century-Crofts, Inc., 1961). In the latter text, the photographs on pages 90-96 are particularly interesting.

21. E. W. Ely, "Simplified Practice in the Gas Appliance Industry," in *American Gas Association, Proceedings 1929*, 781.

22. See *American Gas Association, Proceedings 1929*, 538A-538N.

23. D. Hale, *Diary*.

24. R. L. Hallock, "Refrigerator Sales Methods and Results," in *American Gas Association, Proceedings 1929*, 456.

25. In fact, American Gas Association data combined with Census of Manufacturing data have been reported, indicating that domestic MG consumption was greater than NG consumption after 1921. J. M. Gould, *Output and Productivity in the Electric and Gas Utilities, 1899-1942* (Cambridge, Massachusetts: University Press), 1946, 153, Table A10.

26. *Historical Statistics, Census*, 826.

27. *Historical Statistics, Census*, 718.

28. H. P. Westcott, *Handbook of Natural Gas* (Erie, Pa: Metric Metal Works, 1920), 623-644.

29. Between 1930 and 1932 deliveries of NG to industrial customers declined by 24 percent. Energy Information Administration, *Natural Gas Annual 1984*, DOE/EIA-0131 (Washington, D.C.: Energy Information Administration 1985), 58.

30. D. Dillard, *Economic Development in the North Atlantic Community* (Englewood Cliffs, N. J.: Prentice Hall, 1967), 575-580. For a different view J. K. Galbraith, *The Great Crash* (Boston: Houghton Mifflin Company, 1961). According to Lough sales of some electric appliances were rising in the late 1920's. W. H. Lough, *High-Level Consumption, Its Behavior; Its Consequences* (New York: Mc Graw-Hill Book Company, Inc., 1935).

31. J. H. Tate, *Keeper of the Flame, The Story of the Atlanta Gas Light Company* (Atlanta, Georgia: Atlanta Gas Light Company, 1985), 80-113.

32. The haste in which the pipeline for delivery of NG into Atlanta was extended by Southern Natural Gas Company is in contrast to the pace at which the pipeline for delivery of NG into Philadelphia was extended. A holding company did not have control over the manufacturing gas utility that sold gas in Atlanta. However, a holding company did have control over the manufacturing gas utility that sold gas in the Philadelphia area. Although both pipelines were planned and completed in the late 1920's and early 1930's, gas customers in Philadelphia were not served NG until the late 1940's and early 1950's, whereas gas customers in Atlanta were served NG in the early 1930s.

33. Tate, *Keeper of the Flame*, 96.

4

THE 1930's AND
THE GREAT DEPRESSION

American's binding interest in the speed, instant gratification, and independence that automobiles offered had not waned much with the onset of the Great Depression. In 1930 several times more NG was processed for gasoline than was transported for use in households.[1] Gasoline marketing was also unencumbered by entrenched competitive businesses in local markets, while the growth in NG markets was stunted in many regions because MG, coal, and fuel oil businesses had established ties with other business interests and civic institutions.

Along with the decline in general business activity and in the number of new dwelling units (see Table 4.1), the growth in the number of new NG customers

Table 4.1

New Customers and New Dwelling Units between 1930 and 1939

	1930	1931	1932	1933	1934	1935	1936	1937	1938	1939
Total customers[a]	5095	6456	6506	6628	6983	7391	8017	8348	8634	8887
New customers[b]	354[d]	1360	50	152	355	408	626	331	286	253
New dwellings[c]	330	254	134	93	126	216	304	332	339	458

Source: New Dwelling Units, Grebler, Blank, and Winnick, *Capital Formation in Residential Real Estate*, 332; Customers, *Minerals Yearbook*, various years.

[a] Total number of residential customers in thousands.

[b] Change in the number of residential natural gas customers between years in thousands or the number of new customers.

[c] New private permanent nonfarm housekeeping dwelling units started in thousands.

[d] It is assumed that commercial customers as a proportion of total residential and commercial customers for 1930 were the same in 1929, when only the aggregate number of residential and commercialcustomers was reported.

declined at the beginning of the 1930's. By 1934, the level of capital formation in real estate and non-farm residential wealth had not returned to 1930 levels, yet the number of new residential customers had.[2] Between 1930 and 1939 the change in number of new customers had exceeded the change in number of new dwelling units by 61 percent.

There was a very large increase in the number of new RNG customers in 1931, largely because the number of RNG customers in Illinois increased from 12,970 customers in 1930 to 943,450 customers in 1931, as gas customers in Chicago began receiving NG service for the first time. The number of customers increased not merely because NG cost less than competitive fuels but because city governments and other institutions governing the development of NG distribution had become more receptive to NG service and because some MG equipment had been fully depreciated.

Another contributing factor to the increase in number of customers in 1931 was additional extensions to the pipeline system from the Monroe-Richland field in Louisiania to Atlanta. Since the utility commission in Georgia had stipulated that NG was to be sold at a relatively low price, the major avenue open for a utility to increase revenues was to increase the number of customers.

The number of customers frequently increased for several years after NG service became available to MG customers in an area for several reasons. It provided a level of service at least equal to MG. NG was generally a much less expensive fuel than MG as indicated by a large decrease in the cost of gas when gas customers switched from MG to NG (see Table 4.2). Thus households had more money for purchasing other goods when they switched to NG.

Table 4.2

Indices of the Cost of Gas to Households in Certain Cities

	1929	1930	1931	1932	1933	1934	1935
Atlanta, Georgia	107	63[a]	63	63	63	63	55
El Paso, Texas	100	58[a]	58	58	58	58	58
Lincoln, Nebraska	97	95	51[a]	51	51	51	51
Macon, Georgia	100	100	67[a]	67	67	67	52
Mobile, Alalabama	108	108	61[a]	61	61	61	61
San Francisco, California	102	79[a]	79	79	79	79	79
Springfield, Illinois	93	93	93	74[a]	67	67	67

Source: M. Ada Beney, *Cost of Living in the United States, 1914-1936*, (New York: National Industrial Conference Board, 1936), 80-81.

Note: All cities are indexed to 1923 prices for a city (1923=100), July prices are used. Indices are rounded.

[a] Changed to natural gas service.

DECLINES IN INCOME AND IN NATURAL GAS CONSUMPTION PER CUSTOMER

The drop in RNG consumption per customer (use/customer) and in personal income per capita in constant dollars (income) during the Great Depression was striking. In Ohio, use/customer fell 8.64 Mcf, or 12.5 percent in 1931 from the 1930 level of 69.23 Mcf (see Table 4.3). Real income in Ohio between 1930 and 1931 fell $41 or 6.6 percent from a level of $661. Thus in percentage terms, the decline in use/customer was twice as great as the decline in income. Use/customer in the United States was not to rise to 1930 levels during the 1930's despite a rise in income to 1930 levels by 1936. Large percentage reductions in income were generally coupled with reductions in use/customer between 1930-1931 and 1931-1932 in key states (see Table 4.3).[3]

Table 4.3

Percentage Change in Consumption per Customer and in Real Income

	Variable	1930-31	1931-32	1930-34	1935-39
Ohio	Use/Customer	-12.5	-9.8	-12.7	-0.2
Ohio	Income	-6.6	-20.4	-4.7	+12.8
Pennsylvania	Use/Customer	-11.8	-7.4	-14.7	-5.9
Pennsylvania	Income	-7.4	-16.3	-15.0	+15.3
West Virginia	Use/Customer	-6.3	-5.6	-29.0	0.0
West Virginia	Income	-4.2	-19.6	-29.0	0.0
Oklahoma	Use/Customer	-8.7	+4.7	-27.7	-1.9
Oklahoma	Income	-10.7	-19.4	-15.3	+13.8
New York	Use/Customer	-8.9	-23.5	-32.8	-2.6
New York	Income	-6.6	-14.2	-18.0	+12.2

Sources: Natural Gas Consumption per customer, *Minerals Yearbook*, various years; Personal Income per Capita, Bureau of Economic Analysis, U.S. Department of Commerce, *Personal Income 1929-1982* (Washington,D.C.: Government Printing Office, 1984), 8; Deflator, Consumer Price Index-All Items, *Historical Statistics, Census*, 211.

Note: Income was deflated by the Consumer Price Index.

The magnitude of the decline in use/customer in the first several years of the Great Depression contributed significantly to the overall reduction in the quality of life. "Many families doubled up in their living quarters, while many others migrated from cities back to the farms. The necessity for economy caused many families to curtail their use of both gas and electricity, and many other households which had been using gas for house heating discontinued it for less expensive fuels."[4]

APPLIANCE AND SERVICE DEMANDS

Households that were already using NG for central space heating during the 1930's were in the upper income classes, and most advertising efforts were directed to these households, whose income had risen significantly during the 1920's.[5] In some areas, however, special efforts were made to sell NG as a space heating fuel to many customers by reducing the cost of NG to space-heating customers by 35% or more. Consequently, "The resulting increase in number of house heating customers from 1927 to 1931 was 684%. ... Because of the fact that 16.7% of all domestic customers in Kansas City were using gas for central house heating in 1931, to which must be added over 170,000 room heaters or nearly two per average domestic consumer, house heating sales matched other domestic gas sales in Kansas City in 1931."[6]

Peoples Gas Light and Coke Company of Chicago began mass marketing gas house-heating equipment in 1931. By July of 1931 more than 10,000 installations were made, which were mostly conversion burners.[7] The utility in Denver was also actively marketing house-heating services since the late 1920's.[8] However, such efforts were exceptional at the beginning of the Depression and were a consequence of particularly innovative and aggressive sales managers of local utilities and appliance retailers.

NG was still more expensive than coal and wood in most places, and sellers of fuel oil competed with NG in markets where both sources of energy were available. Fuel oil was preferred by many families who considered it a safer and more reliable source of energy[9] than NG since major accidents involving NG and curtailments in gas service were still occurring. On March 18, 1937 in New London, Texas, 294 people were killed when the Consolidated Public School was destroyed by an explosion of NG piped in for heating purposes.[10] Communities in Ohio, Kansas, and other states had experienced reductions or curtailments in NG service.

Table 4.4

Source of Lighting in American Homes between 1900 and 1940 (in percentages)

	1900	1910	1920	1930	1940
Kerosene	88				
Gas	9	85	65	32	21
Electricity	3	15	35	68	79

Source: Stanley Lebergott, *The American Economy, Income, Wealth, And Want* (Princeton: Princeton University Press, 1976), 279.

A significant number of households still used gas for lighting during the 1930's (see Table 4.4); and despite concerns about safety and reliability, the number of gas-fired warm air furnaces shipped consistently increased in the closing years of the Great Depression from 21,300 in 1936 to 55,500 in 1939. A total of 141,800 furnaces and 121,000 gas conversion burners were shipped during these years.[11] However, shipments of furnaces and burners still amounted to about 2 percent of the total number of customers in 1939.

The number of gas-fired automatic water heaters being shipped near the end of the Depression was increasing. The number of water heaters shipped was 269,000 in 1936; 423,000 in 1937; 387,000 in 1938; and 499,000 in 1939, totaling 1,578,000 heaters or 19 percent of the total number of customers in 1939.[12]

CHANGES IN NATURAL GAS PRICE

Unfortunately, RNG prices in constant dollars were rising dramatically during the early years of the 1930's, as incomes were falling. The increase in price varied by state, with prices in New York changing by more in percentage terms than in any other key RNG state (see Table 4.5). Thus, some of the real income gained by gas customers when they switched from MG to NG eroded as real prices increased.

Table 4.5

Average Residential Natural Gas Prices in Key States (in constant 1930 dollars and in cents per 1000 cubic feet)

	1930	1931	1932	1933	1934	1935	1936	1937	1938	1939
Ohio	64	70	77	80	75	73	72	69	71	78
Pennsylvania	62	68	78	81	77	76	75	71	73	69
West Virginia	38	41	46	48	46	45	45	42	43	44
Oklahoma	54	57	55	58	56	55	55	53	54	55
New York	68	76	88	99	97	101	98	96	98	99
United States	71	78	91	95	93	91	88	86	88	88
Index	100	91	82	78	80	82	83	86	84	83

Sources: Natural Gas Prices, *Minerals Yearbook*, various years; Index, Consumer Price Index-All Items, *Historical Statistics, Census*, 211.

Among the key states the major producing states of Oklahoma and West Virginia experienced the least variability in price. In several states the average real price of NG in the 1930's was almost double its 1910 value. Accordingly, the average level of use per customer during the 1930's was considerably less than it was in the 1910's (see Table 4.6); in fact, it reached its lowest level ever during the 1930's.

Table 4.6

Average Demand and Average Price in Key States by Decade

	Quantity[a]			Price[b]		
	1910-19	1920-29	1930-39	1910-19	1920-29	1930-39
West Virginia	158	147	105	16	16	24
Oklahoma	141	123	85	16	25	31
Ohio	102	82	53	25	29	41
Pennsylvania	125	100	58	23	29	41
New York	115	77	43	27	31	51

Sources: quantity and price, *Minerals Resources or Minerals Yearbook*, various years; Deflator, Consumer Price Index-All Items, *Historical Statistics, Census*, 211.

[a] In thousand cubic feet.

[b] In cents per thousand cubic feet and in constant 1910 dollars.

THE INDUSTRY'S RESPONSE TO THE DECLINE IN DEMAND PER CUSTOMER

In response to the reduction in use per customer and to the competitive threat posed by electrical energy and by the many consumer goods readily available in markets,[13] the gas industry experimented with a variety of approaches for selling more gas per customer and for gaining more new customers as the Depression continued.[14] These approaches were: evaluating the efficiency of a wide variety of NG furnaces and appliances and disseminating this information; providing customers with information about the effect of insulation on energy use; offering maintenance services to gas customers to improve appliance efficiency; experimenting with differences in rates for different appliances to encourage use; joining forces with appliance dealers and plumbers to sell more appliances; selectively putting on cooking and other appliance demonstrations for high school students and new wives; preparing information on appliances specifically geared to builders and architects; developing new appliances that would be affordable to lower income households; providing traveling cooking schools; and advertising the economic gains from home cooking. The industry even evaluated and published information on how much gas was required to can vegetables and fruits and to cook meals. For example, it was reported that a meal consisting of stuffed pork chops with dressing, grilled fruit halves, scalloped baked squash and apple crisp would require 46 1/6 cubic feet of MG.[15]

One of the more interesting attempts made by the gas industry to understand its market was a detailed personal-interview survey of 1,000 gas and electric customers of the Rochester Gas & Electric Company. The topics addressed by the survey were considered to be of great importance to the gas industry. Consequently, it was supported by the the AGA[16] and conducted by an independent research organization.[17]

The survey information indicated that gas ranges were considered qualitatively much better in terms of speed of cooking and length of appliance lifetime but that electric ranges were considered much better in terms of cleanliness, automatic controls, appearances, and extra features. Of those customers who had gas water heaters, 42 percent used their gas water heaters only during the summer. During the winter they still relied on the heat generated from a stove, furnace, or boiler to raise the temperature of water for cleaning, bathing and other purposes. Fifty percent had an ice refrigerator and 5 percent had no refrigerator. For purposes of house heating, 38 percent considered gas to be cleaner than oil and 15 percent considered oil to be cleaner than gas and 40 percent didn't know which to consider cleaner. Of those few customers who were able and willing to compare refrigerators, electric refrigerators were identified as better in terms of speed of freezing, appearance, and extra features, and gas was considered better in terms of silence. Most interviewees were unable to compare the two appliances.

CHANGES IN CONSUMER ATTITUDES

During the 1930's, while the average consumer was responding to reductions in real income and to increases in the real price of NG by reducing the amount of NG used within the household, consumer groups and their representatives were responding at public meetings either to perceived inequities in the price paid by households for NG or to their inability to obtain NG even though there was a well-known surplus of NG in the southwestern United States. These meetings culminated in hearings before the Congress at which legislative proposals were set forth.

At these meetings it was not only the price paid by RNG customers that was of importance to consumer representatives, it was also the price that RNG customers paid relative to the price that industrial customers paid. The discussions at congressional hearings indicated that the equity issue of the relative price of NG to residential and industrial customers was of great concern to consumers and consumer groups.

There was also a response to the power of large gas distribution companies. It was at times similar to the democratic response to the power of trusts, as discussed by Piott[18] for an earlier time period. In fact, a particular natural gas company, the Columbia System, was referred to as the Evil Empire at the congressional hearings.

CONTROL OF MARKETS BY SEVERAL COMPANIES

The iron-fisted control of large gas companies over the distribution system for NG within the United States resulted in communities being unable to obtain NG at prices related to the low prices at which burgeoning pipeline and producing companies were willing to sell them NG from the recently developed NG producing Panhandle region of Texas. As much as 50 percent of the NG was being wasted in the Panhandle because of the control of the large gas companies over market distribution. This region contained more NG reserves than any other region of the United States.

A company, which was to become known as the Panhandle Eastern Company,[19] was formed in 1929 to supply NG to northern and eastern markets from the Panhandle of Texas. Money was raised for pipeline construction by selling stock to the public rather than by obtaining money from internal sources of the gas and related industries. Internal or private sources were the standard method for raising funds for such projects. From the start the project was a threat to entrenched industries. When arrangements were made to sell gas to prospective customers in such cities as Omaha, Minneapolis, Kansas City, St. Paul, and St. Louis, they were stopped by a wide variety of schemes cooked up by these businesses. These schemes ranged from direct threats to the manipulation of the stock of the company. These cities were considered to be part of the territory of an existing pipeline company and, thus Panhandle's initiatives were viewed as an invasion of territory. The companies supporting these schemes included Cities Service, Standard Oil of New Jersey, and Electric Bond and Share.

By the middle of 1931, the Panhandle pipeline system was extended to Indianapolis, Indiana. It cut across the United States and had the capability of serving such cities as Chicago and Detroit and, most importantly, the many cities served by the highly developed pipeline grid in Ohio and western Pennsylvania. The entrepreneurs of Panhandle Eastern, having worked hard at buying up leases and obtaining right-of-ways, had planned to put in several pipelines along the right-of-way. Instead, the pipeline arrived at Indianapolis within the territory of the Columbia System with much excess capacity. Because of financial difficulties related to its inability to sell its gas, the company was taken over entirely by the Columbia System within a few years. Not until 1936 was the pipeline's capacity able to be nearly fully utilized when NG from the pipeline was delivered to customers in the city of Detroit.

PRICE SETTING BEHAVIOR BY MAJOR COMPANIES AND OTHER PRICE ISSUES

Several particular price issues were consistently discussed at the Congressional hearings prior to the passage of the Natural Gas Act of 1938 (NGA).[20] One issue was the unexplained disparity between the average price paid by one group of consumers and another group of consumers when the cost of serving both groups appeared to be the same. In particular it appeared that utilities were charging industrial customers a price for NG that was not based primarily on cost of service considerations. On the other hand residential customers appeared in many instances to be paying a price that was significantly higher than the cost of the service. The high price was frequently viewed as a consequence of the inability of domestic customers to conserve much on their use of NG or to switch readily to an alternative fuel in the short run. Other customers, both residential and industrial, were thought to be charged a high price solely because of their relatively small size and their inability to exert any institutional influence on the price that gas businesses charged them for NG. For example, businessmen could use other business associates, who may have been on the same board of directors as the executives of utilities, as middlemen in their negotiation with gas utilities. It was also claimed that when a majority of the households were served NG within a region, the price rose.

The comments at the hearings indicate that many participants believed that gas businesses were willing to subsidize industrial customers at the expense of residential customers because some portion of the income gained by the industrial customer as a consequence of the lowering of the cost of gas to the industrial customer would be returned to the owners of the gas businesses in the following forms: stock dividends, discounts on equipment and services received from the industrial plant, and other less obvious forms of remuneration, such as electing or designating managers of gas businesses to the boards of directors of large industrial establishments.

There were also good theoretical reasons for industrial customers paying less. By this time corporate planners not only understood that by reducing the price of NG and by adding industrial customers they were able to reduce the average cost of NG to all customers, but they also had some idea of the magnitude of the reduction in cost from such changes in operation.

The demand for RNG tended to be highly seasonal. Average monthly demand was several times as large in the winter as in the summer. Average monthly demand for NG by many industrial customers tended to be relatively flat throughout the year. Thus, the addition of industrial customers would flatten out the overall average monthly demand. Because of the large capital or fixed cost involved in gas transmission, this would tend to reduce the average cost of serving all customers. Experience gained in operating pipeline systems enabled engineers to gauge the magnitude of this reduction.

As long as gas businesses were operating at less than full capacity, and as long as the price NG businesses were receiving for NG from industrial customers was greater than the variable cost of serving the additional NG to the industrial customer, the cost of NG to all customers should have declined as additional industrial customers were served NG. Thus the reduced cost should have led to residential customers having to pay less for NG. This was the typical argument presented for charging industrial customers a price equal to the industrial price of coal, the "competitive" fuel.

Some industrial customers were also charged a low price because they were the first customers of a pipeline company in a new region. Pipeline companies needed the large volume purchases of the industrial customer initially to repay some of the huge debts incurred in building the pipeline system. However, industrial customers continued to receive huge discounts on NG even after these debts were paid.

The consumer representatives understood these arguments but they also thought that, because NG was an exhaustible resource, the large amounts of NG gas used by industrial customers would tend to increase the cost of NG over time. In fact, depletion of wells was frequently offered as an explanation for the steady rise in the real cost of NG since the First World War.[21]

Periods of shortage because the gas industry did not drill enough new wells were still followed by periods of surplus because the gas industry discovered more NG than expected. In addition, extensions of the pipeline system to new wells, as particular old wells were depleted, reduced the average cost of NG service if the new wells, found in more remote regions under better controlled conditions, with improved exploration, drilling, and transportation technologies, were more prolific.

The consumer representatives also understood that the coal resource base, while also finite and depletable, was much larger. Thus, any incremental costs from depletion would be much smaller and would occur at a much later date for coal than for NG. If industrial customers purchased and used more coal and less NG they would, in many instances, be using an inferior source of energy for some of their processes. RNG customers would also be paying a higher price for NG in some instances. However, any increase in the average cost of NG from depletion would occur at a much slower rate and would be more predictable.

The consumer representatives perceived that, if large industrial customers were indiscrimintely served as much NG as they demanded, short-term shortages were also likely to occur more frequently. For example, if the NG industry did not find enough new wells and did not make necessary improvements to the distribution system during a period of industrial expansion, shortages were more likely to occur from the depletion of available wells and from technical problems in distributing the NG during a cold winter because of inadequate pressure.

The consumer representatives seemed to fear that by selling industrial customers as much NG as they demanded at all times at a low price there would be a recurrence of the problems experienced in Indiana. In Indiana relatively new equipment was unable to be used because of actual depletion of the resource base and lack of a pipeline system capable of acquiring NG at sufficient pressure from other regions. Thus, the consumer representatives thought that the economic welfare of residential customers might be much better served if more coal and less NG was consumed by the industrial sector.

The cost of NG to residential customers relative to industrial customers between 1930 and 1940 did not change in most states (see Table 4.7); it only increased dramatically in New York, Oklahoma, and Texas; it even declined in Kansas and in Louisiana. Therefore, it does not appear that a general increase in the price disparity between the residential and industrial customers during the 1930's could have prompted action by consumers and their representatives. Instead, actions were apparently prompted by a continuation during the 1930's of the relative prices that were established in the 1920's when the economic condition of consumers had worsened.

Table 4.7

Residential and Industrial Price by State (cents per 1000 cubic feet)

	1930			1940		
	Residential	Industrial	Ratio[a]	Residential	Industrial	Ratio[a]
New York	68	54	1.3	82	20	4.1
Ohio	64	42	1.5	64	34	1.8
West Virginia	38	25	1.7	36	20	1.8
Pennsylvania	62	35	1.8	57	29	1.9
Kentucky	49	28	2.1	50	26	1.9
Kansas	66	16	4.2	34	12	2.8
Arkansas	52	11	4.3	55	11	5.0
Missouri	98	19	4.9	87	17	5.1
Oklahoma	54	8	5.4	46	5	9.2
Texas	80	9	8.0	73	4	18.2
California	90	10	9.0	78	11	7.1
Louisiana	64	6	10.7	40	8	5.0

Sources: *Minerals Yearbook*, 1930, 466, 472; *Mineral Yearbook*, 1940, 472,1143-1144.
[a] Ratio is the ratio of the residential price to the industrial price.

A price disparity which definitely did motivate citizens to action, however, was the difference in the cost of MG, which was largely used by residential customers, and NG, which was largely used by industrial customers (see Table 4.8). Members of the Cities Alliance, who represented consumers from the north central United States at the natural gas hearings, were very aware of the low cost of NG to industrial customers in the southern states and the high cost of MG to residential customers in their states.

Table 4.8

Average Price of Manufactured Gas and Natural Gas by Sector (per million Btu)

Year	Manufactured Gas				Natural Gas		
	Residential	Industrial	Ratio[a]		Residential	Industrial	Ratio[a]
1932	$2.18	$1.36	1.60		$0.66	$0.16	4.13
1933	$2.15	$1.26	1.71		$0.69	$0.15	4.60
1934	$2.11	$1.15	1.83		$0.72	$0.15	4.80
1935	$2.09	$1.08	1.94		$0.71	$0.15	4.73
1936	$2.06	$1.11	1.86		$0.69	$0.16	4.31
1937	$2.09	$1.03	2.03		$0.70	$0.16	4.38
1938	$2.04	$1.23	1.66		$0.66	$0.15	4.40
1939	$2.00	$1.16	1.72		$0.68	$0.16	4.25

Source: American Gas Association, *Historical Statistics of the Gas Utility Industry* (New York: American Gas Association, 1965), Table 90 Table 92, Table 110, and Table 112.
   [a] Ratio is the ratio of the residential price to the industrial price.

Existing Investment in Manufactured Gas Industry

MG was a widely used, but a relatively expensive, source of energy, which began to decline in importance in domestic markets during the 1930's, but did not begin a major decline until after the Second World War. At the time of the Great Depression, major physical and human capital investment still supported the industry in many towns and cities.

Trained personnel were required to successfully run the operation of a MG plant, which not only supplied gas for lighting, cooking, and other needs of domestic, commercial, and industrial customers but also produced such other products as benzol (for use in dyes and paints), ammonia, resins, and tars. Gas manufacturers (utilities) obtained these chemicals and related products when they processed coal, oil, and other material into a gaseous form for transport to heating and lighting customers. In fact, one of the benefits to society, pointed out by the gas industry, of households switching from coal to MG was that these products were no longer lost to society as a consequence of burning coal in the household but were captured by the gas industry and sold to society when coal was processed into a gaseous form at the gas plant.

The gas plant was more like a chemical plant than like a modern gas utility. For example, tuluol, which was used to manufacture the explosive trinitrotoluene

(TNT), was obtained as a by-product of gas production. Many of the derivatives from gas production ended up as "fertilizer near a South Jersey apple tree, as lacquer in the paint job on that shiny new Ford, as Bakelite in keys of typewriters".[22]

There were half as many MG lines as NG lines at the end of 1936. Moreover, the miles of MG lines exceeded the miles of NG lines in twenty three states in 1932. The ten states with the most miles of MG lines had more than 90,000 miles of gas lines (see Table 4.9). These states, except for Florida, were northern states with high per-capita income, cold winters, and, for the most part, a limited supply of inexpensive, readily available, alternative energy resources. Some of these states, such as New York, Illinois, and Michigan, were states with a growing population, and states with the largest populations in the United States. Thus, these states represented potentially important RNG markets.

Table 4.9

Manufactured Gas Lines in Top Ten Manufacturing Gas States in 1936

State	MG	NG	Total	MG/Total
New York	14,154	3,150	17,304	82%
Illinois	10,610	1,760	12,370	86%
Massachusetts	8,615	——	8,615	100%
New Jersey	8,534	——	8,534	100%
Michigan	5,387	3,600	8,987	60%
Indiana	4,816	1,480	6,296	76%
Wisconsin	4,227	——	4,227	100%
Connecticut	2,681	——	2,681	100%
Minnesota	2,106	400	2,506	84%
Florida	2,072	60	2,132	97%
United States	90,676	179,660	270,336	33%

Source: Temporary National Economic Committee, United States Congress, 76th Congress, 3rd Session, *Investigation of Concentration of Economic Power*, (Washington, D.C.: U.S. Government Printing Office, 1940), 213.
Note: NG is natural gas; MG is manufactured gas and MG/Total is manufactured gas lines relative to Total gas lines lines expressed as a percentage.

Revenues received by utilities from the sale of MG at the beginning of the 1930's (see Table 4.10) exceeded the revenues received from sales of NG even though the volume of NG sold was more than twice as great as the volume of MG sold. When compared to the average industrial customer, the average residential customer paid almost twice as much for MG and more than four times as much for NG. Thus, the savings possible for residential customers using MG, who were able to obtain NG near the industrial rate, were phenomenal. It was such differences in cost which, in part, motivated the formation of the Cities Alliance. The Alliance advocated complete freedom for institutions representing domestic consumers to purchase NG directly from a NG production company and to arrange either for transportation of the NG by an existing pipeline or for construction of a new pipeline.

Table 4.10

Revenues from Manufactured, Natural, and All Gas (in thousand dollars)

	Manufactured Gas		Natural Gas		All Gas	
	Total	Pctg[a]	Total	Pctg[b]	Total	Pctg[c]
1932	359,884	82%	300,792	64%	660,676	54%
1933	320,953	82%	291,878	62%	612,831	52%
1934	318,878	81%	315,638	58%	634,516	50%
1935	310,619	80%	341,949	57%	652,568	48%
1936	303,671	80%	387,652	55%	691,323	44%
1937	292,458	79%	430,717	55%	723,175	40%
1938	291,105	77%	406,352	58%	697,457	42%
1939	293,138	76%	438,637	57%	731,775	40%

Source: American Gas Association, *Historical Statistics*, 1965, Table 110, Table 112.
[a] Percentage of residential manufactured gas revenues to total manufactured gas revenues.
[b] Percentage of residential natural gas revenues to total natural gas revenues.
[c] Percentage of manufactured gas revenues to total gas revenues.

MG revenues from residential customers remained a sizable proportion of revenues received from all customer classes during the 1930's. Although MG revenues received from residential customers declined during the 1930's as many of these customers began to switch to NG, the decline in revenues between years was rather modest except for the change between 1932 and 1933 (see Table 4.10).

Vested Interests of the Coal and Related Industries

In the 1930's the coal industry was represented by a vocal union gaining wage and other concessions both from the owners of mines and from the national government. The industry was represented by businesses that were becoming increasingly mechanized.[23] Mechanical loading rather than hand loading from underground mines was becoming increasingly common, both in bituminous and in anthracite mines. In 1928, 3.4 percent of anthracite coal and 4.5 percent of bituminous coal was mechanically loaded. Ten years later 27 percent each of anthracite coal and of bituminous coal was mechanically loaded. In addition, approximately 90 percent of the bituminous coal mined was mechanically cut by the end of the 1930's. Coal miners were being laid off because of this mechanization and because of the decline in overall demand.

The coal workers and the mines were in nonurban areas. Moreover, the industry was made up of many small operators. Consequently, although particular dealers did advertise the disadvantage of switching from coal to NG for residential and commercial customers,[24] the industry did not invest much of its resources in protecting coal markets from the development of RNG markets, and the coal industry was not well represented at the congressional hearings that resulted in the NGA. Coal continued to remain important in residential markets during the 1930's either because households did not have the income necessary to switch to an alternative fuel, an alternative fuel other than wood was not available or because of relative prices. Coal did not continue to remain important because of steps taken by the industry to prevent the erosion of its market.

The residential sector was also not the major consuming sector for coal. This may explain the appparent lack of interest by the industry and the coal unions in the threat posed by the further development of the NG transmission system in the North.

MG utilities, however, did have an indirect role in determining the rate at which the coal industry declined in importance for residential energy markets in the 1930's. These utilities purchased coal, many under long-term contracts, from coal producers and coal distributors and had developed business relationships with these companies. They also developed business relationships with the companies that purchased the additional products from coal gasification, and with the large manufacturing plants that used coke ovens. These firms used coal to produce coke, which in turn produced coke oven gas, which they sold to gas utilities.

Coke oven gas was mixed with other forms of gas such as NG and then sold to residential customers. The use of this gas was common in such cities as St. Louis, Cincinnati, Baltimore, and Philadelphia. These cities still did not have full NG service during the 1930's, even though pipelines of sufficient capacity may have extended to within a few miles of them.

The growth of RNG markets into the regions served by coal was made difficult by entrenched MG interests and by the limited supply of underground NG storage. It was also recognized by many that major changes in the regulatory environment would be required in order to effectively promote further development of RNG markets.

Actions Leading up to the Passage of the Natural Gas Act of 1938

The Federal Trade Commission (FTC) prepared a series of reports on the gas industry based on hearings and other sources of data during the 1930's.[25] The studies grew out of a massive investigation initiated in 1928 of the gas and electric industries. After Franklin Delano Roosevelt (FDR) was elected president in 1932, not only were these studies continued, but commissioners of the FTC who were known to have a strong industry bias were fired, and many new people were hired.

Thomas Corcoran, one of FDR's most trusted lieutenants, who was instrumental in the passage of the Public Utility Holding Company Act, had the time of his public service life preparing testimony, exerting political pressure, and helping to fill vacancies on task forces at government agencies. He filled these vacancies with professional staff who shared his enthusiasm for the Second New Deal and a more active role for the federal government as an institutional antidote to the power of large corporations, especially holding companies. A primary mission was to reduce the economic and political power of holding companies.

Holding companies were even viewed by some as a direct threat to the democratic character of American society. Holding companies not only controlled many gas and electric utilities in many parts of the country but also frequently controlled NG production and distribution companies as well as coal- and oil-producing industries.

Table 4.11

Control of Natural Gas Industry by Rockefeller and Morgan Interests

	Produced	Trunk	Transported
Rockefeller Interests (RI)	13.6%	12.3%	16.8%
Companies Connected with RI	9.4%	15.8%	33.1%
Morgan Interests (MI)	11.1%	34.8%	23.3%
Companies Connected with MI	2.7%	17.1%	14.4%
Total[a]	36.2%	80.0%	87.7%
Total U.S. Amount[b]	1,771	52,000	471

Source: Temporary National Economic Committee, *Investigation of Economic Power*, 99-100.

Note: Produced is the the amount of natural gas produced from wells. Trunk is the miles of trunk or long distance pipeline. Transported is the amount of natural gas transported across state lines.

[a] As a percentage of U.S total.

[b] Produced and transported amounts are in billion cubic feet. Trunk amount is actual mileage of pipeline in the United States.

As part of the FTC investigation, company records (when available) were examined in detail. The amount of control over the NG industry exercised by a few corporate entities (see Table 4.11) was well presented in particular FTC reports.

The FTC identified sixteen specific evils of the monopolistic structure of the gas industry. Eight of these sixteen items were particularly relevant to the cost of NG, the availablity of NG, and other aspects of the economic welfare of households. These eight were: excessive cost of NG production because of excessive competition in drilling wells; costly struggles between rival NG interests to conquer or defend territories of distribution; excessive and inequitable variations in city gate rates for NG among different localities; excessive profits in many NG sales between affiliated companies; exploitation of subsidiaries of NG companies through fees for construction, management, and promotion; exaction of excessive bonuses or commissions by investment bankers in connection with financial transactions with NG gas companies in certain instances; exaction of excessive bonuses or commissions by officials of certain companies in connection with sales and construction of properties; and misrepresentation of financial conditions, investment, and earnings of some NG operating and holding companies.

The FTC findings recommended greater federal control of the gas industry. The FTC inquiry provided support for the passage of the Public Utility Holding Company Act of 1935. The inquiry also supported the enlargement of the jurisdiction and functions of the Federal Power Commission (FPC) through the Federal Power Act of 1935 and findings from the study were also used as support for the passage of the Securities Act of 1933. Seperate titles of the Federal Power Act also created the Securities and Exchange Commission.

In 1936 representative Clarence Lea of California introduced a bill in the House that was essentially Title III of the original Public Utility Holding Company Act of 1935 and which was not included in the final Act. Hearings on this bill, HR 11662, which was described as a bill to regulate the transportation and sale of NG in interstate commerce, were begun on April 2, 1936. The bill proposed to give the FTC regulatory control over those price and service aspects of the transportation of NG between states by pipeline companies which were not under the control of state commissions. The bill stipulated nothing about the construction of pipelines. Thus, there would have been no control over the number of pipelines entering a market if it had become law. Action on HR 11662 did not proceed beyond hearings during the election year of 1936, when FDR defeated the Republican, Al Landon, in a landslide victory.

Prior to 1936 when the FTC was studying the gas industry, and during 1936 when Congressman Lea was conducting hearings on HR 11662 for the regulation of interstate NG pipelines, representatives of urban communities continued to lobby for federal control of the gas industry. These representatives could be relied upon to respond at congressional hearings or other public forums to questions about the effect of the actions of utilities, NG pipeline companies, and producers on the welfare of consumers. Their responses could vary from verbal support for any group opposed to the power of the NG industry to the presentation of detailed price statistics.

The price statistics were presented by the representatives of the urban communities to support the claim that the NG industry frequently charged different prices in different communities that did not reflect differences in cost of serving customers in these communities. The price statistics were also offered as evidence that NG companies frequently charged a much higher price to an affiliated company outside of the borders of a state than the price that the NG companies had paid. The affiliated company was then able to pass this cost off to residential customers in another state because state utility commissions only had jurisdiction over sales within a state. Thus, they were able to circumvent the regulatory authority of state commissions over price by this relatively simple procedure.

The Cities Alliance, representing about 100 midwestern urban areas, was a mass movement that was not fettered by parochial interests. The city representatives, many of whom came from major coal mining states, also did not see themselves as allied with rural consumers in their state, many of whom were coal miners and would be affected by any legislation that increased the availability of NG.

Since the Cities Alliance had direct channnels to the legislative process, a bill (HR5711), drawn up by members of the Alliance, was introduced into the seventy-fifth Congress in 1937 by Representative Robert Crosser of Ohio and Senator Prentiss Brown of Michigan. An important stipulation of this bill was that pipelines could elect either to be treated as common carriers or, if they purchased NG directly, to not demonstrate undue preference for any candidate supplier of NG to them in a field. The bill would also have required pipeline companies to provide NG service to communities that were able to extend a service line to the main pipeline.

Crosser, a member of the commerce committee, deferred to the committee's chairman Clarence Lea of California, who sponsored the more moderate HR 4008. Hearings on both bills were conducted in early 1937. The Cities Alliance objected to the portion of the bill that would make the cities reliant on one supplier. Their argument was that allowing cities to take an active part in the acquisition of NG would provide needed income to southwestern producers, who had a large surplus of NG, and would also reduce the flaring of NG in the production of oil and the use of NG in carbon black manufacture, thus preserving the heat value of NG for subsequent sale in residential markets. Consumers were willing to pay a price that included the cost of producing NG plus a processing and transportation cost and enough extra to make it economically worthwhile for all participants. The NG industry, however, did not want its role as a buyer and seller subverted by an aggregate of consumers, such as a municipal authority, buying directly from producers.

The hearings on the bill attracted much less attention from the public, the Roosevelt administration, and the congress than did the Public Utility Holding Company Act. FDR had always taken a keen interest in electric power issues, and whereas the price, control, and availability of electrical energy was an issue in every state, NG was primarily an issue in the states bordering the Great Lakes.

Electricity was also viewed as the source of energy that would change the American way of life. NG was still primarily viewed by many only as a relatively inexpensive and clean source of energy to replace coal and wood. Moreover, local governments with their many municipal electric utilities and the national government with the Tennessee Valley Authority and the rural electrification program were much more involved in electric power issues.

THE NATURAL GAS ACT OF 1938

Whereas the Holding Company Act was in the courts until late in the 1940's, the NGA hearings of 1937 and 1938 resulted in a compromise, and a bill was signed into law in June of 1938 after being passed by the House and Senate in 1937. Interstate pipelines would be regulated, but the monopoly position of a pipeline company in a service area would be legitimized. Republican Representative Charles Hallech of Indiana argued effectively that if pipelines were to be regulated they should also be protected from cutthroat competition.

Chairman Lea of the commerce committee, in presiding over the hearings leading to the enactment of the NGA, stated that modern regulation favored a certificate of convenience and necessity; this meant that a strict procedure would protect the interests of all parties, before a community could obtain NG from any source other than the source from which it had initially obtained it. On the other hand, pipelines could not abandon service to a community.

The reason for regulating the gas industry was not to discourage the distribution of NG to carbon black and natural-gas gasoline processing plants, where the heating value of NG was largely ignored, nor to stop the flaring of gas because of a lack of a market to northern cities. The major reason for regulating NG pipelines was to protect capital investments of businesses from competitive forces. As a consequence of this protection, the government would also have the regulatory responsibility to protect households from curtailments in service after they had been hooked up to a NG line and to prevent pipeline companies from passing on excessive costs incurred in transactions between affiliated companies outside the jurisdiction of a state where the NG was delivered. But an institutional setting was not created which would have supported or required agents for households from a state or federally funded consumer organization to seek out least cost NG.

The final positions stated in the hearings, which would become the key points in the NGA, were straightforward and not much different from some of the major positions of the competing interests. The Federal Power Commission (FPC) would serve in an oversight role for the activities of the interstate pipeline industry, but would not direct activities or encourage market forces to operate. The pipelines were to be protected from unbridled competition and would, effectively, be assured markets at prescribed prices for NG. The final position did not imply any dramatic change in the economic welfare of any group, either in the short term or in the long term. Thus, the Lea bill was agreeable to all parties and passed the House and Senate with only minor amendments. The FPC assumed control over the price charged for NG by interstate pipelines and encouraged the establishment of just and reasonable rates without undue price discrimination among customer classes. The FPC also assumed control over extension and service abandonment by pipelines and over entry of new pipelines in end-use markets.

RESIDENTIAL MARKETS FOR NATURAL GAS AT THE CLOSE OF THE 1930's

By the end of the 1930's, the NG industry and the RNG market had developed much beyond the role of supplying end-users directly from a well, a role which had characterized its earliest years. Underground NG storage, liquefied NG (LNG), and long-distance pipelines had been shown to be operational under varying conditions. Thus, at the close of the 1930's NG was being moved much greater distances, and larger amounts of NG were being stored for longer periods of time than in 1910. Although the Cities Alliance had not been effective in creating a freer market for NG in which elected representatives of end-users would deal directly with suppliers of NG to obtain NG, it had posed a recognized threat to the NG industry and had helped to create an institutional setting in which more efficient exchanges were likely to occur.

The map presented in Figure 4.1 summarizes fairly well the development of the NG industry as of 1937. The dark, relatively dense lines near the Pennsylvania, West Virginia, and Ohio borders reveal where the NG industry began and where an interconnected NG pipeline system was first developed. In the area near the Kansas, Oklahoma, and Missouri borders, another early region developed. The map also indicates relatively isolated pipeline systems in Wyoming and in California. Pipelines were extended outwards from the south-central United States generally toward the north and east during the late 1920's and early 1930's, until they came close to or reached the major cities of the north. A pipeline system was also extended from the Appalachian region towards more eastern cities. These cities, however, would not be served with NG until several years after completion of the pipeline system. By 1937 NG pipelines had been extended into most states except the New England and northwestern states, the Carolinas, and Nevada. In 1940 NG was available to households in most major states.

Figure 4.1 The Early Natural Gas Transportation System

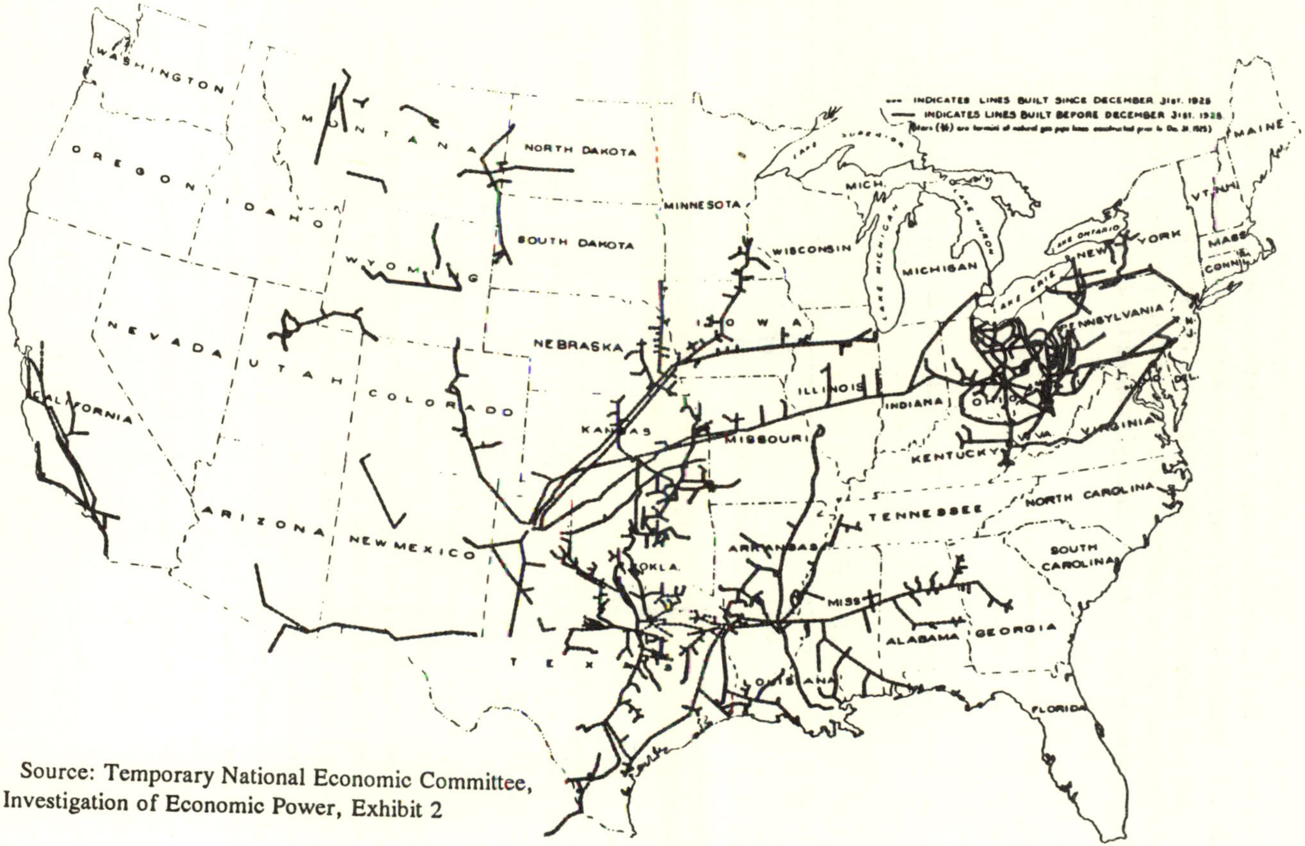

Source: Temporary National Economic Committee,
Investigation of Economic Power, Exhibit 2

A Close Look at Gas and Other Fuel Use in Households in 1940

By 1940, if NG was not used in a state, MG was used, if only for cooking. Gas was the principal source of energy for cooking in almost half the states in 1940. In that year the gas used in households in any one state was almost equally likely to be NG (see Table 4.12) as MG (see Table 4.13). The other energy needs within the household, however, were largely satisfied by using coal and wood. Grime

Table 4.12

Cooking and Space-Heating Fuels in 1940 in Natural Gas States (as percentage of total households)

	COOKING					SPACE HEATING				
	Gas	Wood	Coal	Ker	Elec	Gas	Coal	Wood	Oil	Ker
CA	81	9	0	3	6	66	1	17	6	2
OK	49	32	2	16	0	45	12	40	0	1
TX	44	32	0	21	2	43	3	46	0	0
OH	68	4	14	7	6	7	88	3	0	0
IL	66	5	18	7	4	2	86	3	7	0
MI	56	13	8	10	13	3	79	9	7	0
KS	48	20	7	22	4	28	41	28	3	0
WV	47	16	34	1	2	30	60	10	0	0
MO	42	29	10	15	4	5	61	29	3	0
MN	40	35	6	8	9	3	51	9	16	0
WY	37	13	42	4	3	26	58	11	2	6
CO	34	8	49	4	5	6	82	11	0	0
IA	34	21	21	15	5	2	71	17	9	0
MT	28	27	33	2	9	14	46	24	3	3
KY	27	31	34	6	2	6	72	21	0	0
SD	17	26	29	18	6	5	65	20	2	0
ND	13	14	52	14	7	2	85	6	4	0
LA	34	50	0	14	1	29	7	55	1	0
AZ	31	44	3	14	7	20	4	43	17	3
NM	24	49	17	8	2	7	24	46	5	0
AK	17	76	1	6	1	15	6	77	0	0
GA	13	68	3	9	6	4	49	55	0	0
TN	11	54	21	6	8	3	56	40	0	0
MS	10	83	1	4	2	9	10	79	0	0

Source: U.S. Bureau of the Census, *Census of Housing 1940*, various pages.

Note: The value 0 indicates less than 0.5%. The percentages may not sum to 100because of roundingand because some households reported other fuel or no fuel. For the column headings; Coal is coal or coke; Ker is kerosine; Elec is electricity; and Oil is fuel oil.

Table 4.13

Cooking and Space-Heating Fuels in 1940 in Manufactured Gas States (as a percentage of total households)

	COOKING					SPACE HEATING				
	Gas	Wood	Coal	Ker	Elec	Gas	Coal	Wood	Oil	Ker
CT	61	6	8	15	7	0	49	7	37	5
NJ	79	3	9	7	2	1	70	2	21	3
MA	58	4	9	21	4	0	56	4	37	8
WI	45	32	4	10	8	0	62	27	9	5
RI	60	3	8	22	4	0	48	4	37	8
DE	55	19	6	15	4	0	56	19	22	1
NY	80	6	7	5	3	3	70	4	16	2
PA	61	5	26	4	4	4	86	3	5	0
IN	43	11	21	18	6	0	87	9	0	0
OR	17	59	1	1	22	2	4	73	15	0
NC	5	67	5	17	6	1	42	54	2	0
SC	5	70	2	17	5	0	36	59	3	2
NH	18	39	6	28	8	0	32	74	25	4
VT	12	52	7	19	9	0	36	48	14	1
ME	11	57	6	16	5	0	29	53	15	1
ID	1	43	34	1	22	0	53	39	3	0
NV	12	33	30	3	20	0	34	28	29	0
MD	60	16	9	10	4	2	20	15	20	4
FL	20	31	0	37	11	3	3	54	6	3
UT	16	8	57	0	19	9	81	6	7	1
AL	11	67	13	5	5	3	40	56	3	0
WA	11	57	6	1	23	0	12	56	22	0
VA	21	46	19	9	6	0	51	40	8	16

Source: U.S. Bureau of the Census, *Census of Housing 1940*, various pages.

Note: The value 0 indicates less than one-half percent. The percentages may not sum to 100 because of rounding and because some households reported other fuel or no fuel. For the column headings; Coal is coal or coke; Ker is kerosine or gasoline; Elec is electricity; and Oil is fuel oil.

and dirt were created when these fuels were handled and stored, and dangerous by-products were vented to the environment when these fuels were burned. Man-created smoke in the air outside the household was as common a sight as the naturally occurring fog had once been.

The convenience, cleanliness, and efficiency of gas as a source of energy was clearly recognized, as expensive MG was used widely for cooking in the economically prosperous north. The less prosperous parts of the United States used coal and wood.

In 1940 more than 2 percent of the dwelling units in the United States did not even have heating equipment and 58 percent had noncentral space-heating equipment; of these, 37 percent used wood. As their primary source of space heat, more dwelling units used wood than oil and gas combined. Fifty-four percent of the dwelling units used coal/coke as their primary source of space heat, and of these, 76 percent or 14.3 million dwelling units had central space heating.

Many of the households in states with the lowest per-capita income, such as Mississippi, still used wood for both space heating and cooking. Most households in the Northeast used coal and had central space heating. More than half of the dwelling units in the southern states of Mississippi, Tennessee, Georgia, Arkansas, Louisiana, Alabama, South Carolina, and North Carolina used wood as the major source of energy for cooking. Wood was also used widely in the upper northeastern states of Maine and Vermont and in the northwestern states of Oregon and Washington. This was a vestige of the nineteenth century when wood was the major fuel for both cooking and for space heating within households. Because of the underdeveloped state of underground NG storage and the high cost of NG relative to coal, and because of available coal space heating equipment, a large proportion of households in such states as Ohio and Michigan still used coal for central heating.

During this time period, two very important, but much different, markets for NG had developed in the states of California and Illinois. California had more NG customers than any other state, and customers in California used NG extensively for cooking and space heating. Illinois had a very large number of customers, but these customers used NG mostly for cooking.

The Market in California

In 1940 California accounted for 38 percent of the housing units in the United States which used gas as their primary source of central space heating. In California a large proportion of the households had gas service and an especially large percentage had gas central heating. This situation contributed to the popular image of a high standard of living available to the average citizen in California.

Since the weather was milder in California than in most states, the average NG customer in California during the 1930's used only about 40 Mcf of NG per year. Because of the relatively moderate variation in weather throughout the year, the level of NG demand throughout the year was also relatively constant in California. Therefore, the need for underground storage to support the RNG market in California was much less than in the northern states. Thus markets could be developed even without a well- developed gas storage system.

There are other reasons for the extensive use of NG in California. Coal and wood were not widely available in many parts of the state. The NG and oil mining industry had been experiencing phenomenal growth since the close of the nineteenth century, and this industry had developed an extensive pipeline system within California. Two large gas companies, which had set out over the years to be the dominant gas companies in California by acquiring other companies, were involved in delivering NG to practically all the RNG customers in the state and were very active in pursuing new customers.[26] Pacific Gas & Electric (PG&E), based in San Francisco, served the northern and central parts of the state. In 1936, PG&E had 7,921 miles of gas mains, 571,000 customers, and sold 53 Bcf of NG. Pacific Lighting Corporation, based in Los Angeles, and its subsidiaries served the southern part of California. For the year ending December 31, 1936, Pacific Lighting and its subsidiaries delivered 40 Bcf to residential and commercial customers alone.

The Market in Illinois

Illinois had begun to emerge as a major RNG market during the 1930's. By 1935, there were more than one million RNG customers in Illinois. Only the states of Ohio and California had more customers. Ohio, Illinois, and California accounted for 45 percent of RNG customers in the United States in 1939.

The majority of the RNG customers in Illinois resided in Chicago, where the winters were windy and cold. During the 1930's the cost of NG in Illinois was high, generally greater than $1.30 per Mcf. Thus, a family that used NG for central space heating and cooking in 1935-1936 would have been spending about $350 or 28 percent of total expenditures of the average family in those years.[27] Consequently, most customers did not use gas for central space heating and the level of use per customer was very low, generally less than 16 Mcf per year.

The major gas company in Chicago, Peoples Gas Light and Coke Company, was actively pursuing new space heating customers in the 1930's. Peoples in 1937 made a detailed market study of its service area.[28] One purpose of this study was to identify the potential market for different gas appliances. Using Bureau of Census and other data, researchers had developed a series of analytic maps of Chicago. These maps showed for each square mile of Chicago the number of single family residences (SFRs), the number of SFRs with a central heating plant, the number of SFRs with income in excess of $2,500 and residential property between $5,000 and $7,500. Prime areas were defined as those areas that contained households with a large proportion of central space-heating units, with high income and with high property values. The four square mile area bounded by Bryn Mawr, Central, Western, and Devon Avenues was such a prime area. Peoples knew that more than 99 percent of the single family residences in this area had central space heating units and that the number of gas central space heating units had increased by 23 percent between 1933 and 1936. Markets, however, would not become fully developed in Chicago and in the rest of Illinois until sufficient amounts of underground NG storage became part of the distribution system.

Underground Storage at the Close of the 1930's

Significant progress was made in developing underground NG storage to support market growth at the close of the 1930's. In 1936 only seventeen underground NG storage reservoirs had been developed in the United States.[29] In 1936, two storage sites each were developed in West Virginia and Ohio for the first time. In the next three years, there were twenty-four new underground NG storage reservoirs developed in the United States from Kansas and Oklahoma to New York. There were thirteen reservoirs developed in Pennsylvania alone.

Underground storage reservoirs had been developed for a number of reasons.[30] Peoples Natural Gas Company in Pennsylvania had used storage to relieve pipeline pressure when it built up. The North Penn Gas Company, also in Pennsylvania, had developed underground NG storage because of the large withdrawals of gas by companies operating in the same area as they did. Thus North Penn withdrew all the NG that they could from such areas and transported it to a storage reservoir where they were the sole owner and operator of the reservoir. In Oklahoma underground storage was used to prevent the waste of residue gas from gasoline plants which were connected to a pipeline system. The supply of the residue gas exceeded the demand for it during low demand periods, and that portion of the gas which could not be utilized in the pipeline system was being wasted in the air. Thus one of the major technological constraints for accelerated growth of RNG markets, the slow pace at which underground storage reservoirs were developed, was overcome near the close of the decade.

NOTES

1. *Minerals Yearbook*, 1931, 457-458. In fact, more NG was processed for gasoline than was consumed by commercial, industrial, and residential customers.

2. The domestic NG price and quantity data presented represents, for the most part, residential customers exclusively. The data presented up to 1930 represents the combined residential/commercial sector of which the residential sector was the major sector, based on results for 1930, in which data for the combined sector and separate sectors were reported.

3. The average level of income in a state, however, was not related to the level of use in the generally expected way. In fact, the state with the highest level of income, New York, was also the state with the lowest level of demand per customer. The states with the lowest level of income, Oklahoma and West Virginia, were the states with the highest level of

demand per customer. There are several possible explanations for this. The level of per-capita income for a state, such as New York, probably did not represent the level of per-capita income for gas customers in New York who resided in the southern and southwestern portions of the state. Moreover, differences in the level of per capita income between states were likely to be associated with differences in the levels of demand between states only if the average customers in the two states used NG for the same set of end uses and if adequate supplies of NG were readily available for all end uses. Thus, higher level of use in West Virginia when compared to New York probably reflected a combination of a more extensive use of NG in West Virginia due to a lower price and the much greater availablity of NG for all end uses, and, possibly, such other differences as types of housing structures and appliance stocks in the two states.

4. Dawson, *Columbia System, A History*, 126.

5. Lee and Passel, *A New Economic View of American History*, 333-361.

6. United States Congress, 76th Congress, 3rd Session, *Investigation of Concentration of Economic Power*, Temporary National Economic Committee (Washington, D.C.: United States Government Printing Office, 1940), 144.

7. Conversion burners were used to convert an existing non-gas central space heating unit to gas, see J. A. Clark, *The Chronological History of the Petroleum and Natural Gas Industries* (Houston: Clark Book Co., 1979).

8. Rose, "Urban Environments," 503-539.

9. Reliability was one of the arguments cited by large gas conglomerates, such as Columbia Gas System, for favoring large interconnected enterprises. One of the major social welfare arguments for the institutional necessity of large gas companies was that because of their size they were able to circumvent shortage problems from depleted wells in one region by supplying needed gas from another source such as a well that had previously been shut-in in another region.

10. Clark, *Chronological History*, 193.

11. Gas Appliance Manufactures Association, Inc., *Statistical Highlights*. September 10, 1975.

12. Gas Appliance Manufactures Association, Inc., *Statistical Release*. September 12, 1975.

13. B. N. Behling, "Competitive Substitutes for Public Utilities Services," *American Economic Review*, 27 (1937):17-31.

14. For details on specific examples, see articles in the Commercial Section of *American Gas Association, Proceedings 1937*.

15. A. M. Bailey, "Substitutes for Cooking Schools," *American Gas Association, Proceedings 1937*, 388.

16. American Gas Association, Rochester Domestic Customer Study, Appendix II, Report of the Rate Structure Committee, *American Gas Association, Proceedings 1937*, 52-75.

17. It is worth noting that a number of quality control checks on the reliability of the data were made. Individual data items were verified by making a second call on the customer whose name was secured on the first call. Information from the survey on appliance saturation for customers was also cross-checked with company records. The interview was made up of thirty seven detailed questions which took up four typed pages.

18. S. L. Piott, *The Anti-Monopoly Persuasion, Popular Resistance to the Rise of Big Business in the Midwest*, (Westport, Connecticut: Greenwood Press, 1985).

19. Federal Trade Commission. *Utility Corporations*, 62-70.

20. The material in this section and the following section draws heavily on the transcripts from two hearings. United States Congress. *Hearings before a Subcommittee of the House, Committee on Interstate and Foreign Commerce on H.R. 11662 (Natural Gas)*, 74th Congress, 2nd Session 1936 and United States Congress, *Hearings before a Subcommittee of the House, Committee on Interstate and Foreign Commerce, Hearings on H.R. 4000, (Natural Gas)*, 75th Congress, 1st Session, 1937.

21. It was never clear whether an author intended depletion to mean a constantly declining rate of production from newly discovered wells, a constantly increasing cost of finding new wells from the population of available wells, or just a declining rate of production from existing wells.

22. J. C. G. Conniff and R. Conniff, *The Energy People, A History of PSE&G* (Newark, New Jersey: Public Service Electric and Gas Company, 1978) 257.

23. Barger and Schurr, *The Mining Industries*, 174, 185.

24. Ibid., 524.

25. Interesting discussions of the FTC reports and of the legislative history of the Natural Gas Act can be found in several publications. Sanders, *The Regulation of Natural Gas*. R. K. Huitt, "National Regulation of the Natural-Gas Industry," in *Public Administration and Policy Formation*, ed. E. S. Redford, (New York: Greenwood Press, 1956), 53-116. W. H. S. Stevens, "The Federal Trade Commission's Contribution to Industrial and Economic Analysis: The Work of the Economic Division," *The George Washington Law Review*, 8 (January-February 1940): 545-580. P. Funigiello, *Toward a National Power Policy*, (Pittsburg: University of Pittsburg Press, 1973). D. J. Libert, "Legislative History of the Natural Gas Act," *The Georgetown Law Journal*, 44 (1956): 695-723. Discussions of utility regulation during the Roosevelt years and the degree of involvement of such individuals as Thomas Corcoran in putting new legislation in place can be found in A. M. Schlesinger Jr., *The Politics of Upheaval*, (Cambridge: The Riverside Press, 1960), 211-241, 302-323, 361-408. M .L. Ramsay, *Pyramids of Power*, (New York: Bobbs Merrill, 1937). J. P. Lash, *Dealers and Dreamers, A New Look at the New Deal*, (New York: Doubleday, 1988), 191-214.

26. Stotz, *History*, 309-330.

27. *Historical Statistics, Census*, 327.

28. Market Analysis for Sales and Rate Making, Appendix 1, General Session, *American Gas Association, Proceedings 1937*, 37-49.

29. J. V. Goodman, *Early History of Underground Gas Storage in the United States, American Gas Association, DMC-62-41*. (New York: American Gas Association, 1962), 2.

30. Symposium - Our Experience with the Underground Storage of Gas, *American Gas Association, Proceedings 1940*, 168-203.

5

THE SECOND WORLD WAR
AND ITS AFTERMATH

As Hitler moved his armies across Europe in 1940 and as Europeans with scientific and engineering training continued to emigrate from war-torn Europe to the United States, Americans could observe one of the most significant dramatizations of business enterprise within the setting of the household - television. This product of American engineering talent would engage the attention of the family in a way that conversation and reading never had. The purchase of a NG central space heating unit would allow for a warm, untroubled environment for the proper viewing of television.

The conditions for obtaining NG at least cost were much improved by 1940. The NG-producing industry, RNG customers, and the watchdogs of the industry in Washington, in state capitals, and in cities had available large amounts of information on such diverse factors as the price of NG in different places, the level of the NG resource base, and the location of different distribution systems. This information, which was a by-product of the enormous amounts of data collected and interpreted by economists, statisticians, and other analysts at government agencies during the Great Depression, would encourage more efficient exchanges of NG. The exchange of price information by telephone by former members of the Cities Alliance living in different cities would, in particular, encourage such exchanges.

By 1941 more income was being distributed towards the lower and middle-income classes. Between 1929 and 1941 the percentage of total income received by the top quintile (20 percent of families fell from 54.4 percent to 48.8 percent but the percentages received by the next two highest quintiles rose from 19.3 percent to 22.3 percent and 13.8 percent to 15.3 percent, respectively.[1] Thus, more consumers were able to afford more of the upper middle-income style of life with extensive gas and electric service that had become fashionable during the second decade of this century.[2]

During the 1940's the price of NG declined; improved refrigerators, ranges, and ovens were being sold to gas customers; and an increasing proportion of NG customers were using NG for space heating. These changes resulted in an increase in the amount of NG consumed per customer.

By 1940 thirty-four states were receiving NG service (see Table 5.1). But, such service was being received by more than 50 percent of the households in only three states. In rural states such as Alabama, South Dakota, Tennessee, Mississippi, Georgia, Iowa, and Arkansas, only a very small proportion of the housing units had access to NG service, and citizens in the great metropolitan areas of Philadelphia and New York City had still not received NG service. As a consequence, the states of Pennsylvania and New York paled in comparison to many of the other states in terms of the proportion of households receiving NG service. In contrast, even though the states of Illinois, California, and Ohio accounted for only 18 percent of the occupied housing units in the United States, they accounted for 44 percent of the RNG customers in the United States since more than half the housing units in each of these highly urbanized states were receiving NG service.

RNG customers in California, Illinois, and Ohio differed greatly. Use per customer in California was about average for the United States, but when compared with the value for heating degree days (HDDs), usage was high. Use per customer in Illinois was particularly low when compared to the level of HDDs.

Whereas California represented an indigenous market, Illinois was importing much of its NG from Kansas, Louisiana, Oklahoma, and Texas. As a consequence the cost of RNG was high in Illinois, and each RNG customer purchased less NG than in any other state. Each customer in Illinois used approximately the amount of NG required to cook food on a gas range. Customers in Ohio received much of their NG from within the state and from the nearby state of West Virginia and paid half as much for NG as customers in Illinois. The average customer in Ohio used approximately the amount of NG that would be required to cook on a gas range and to heat water with an automatic water heater.

Although the average RNG customer in Pennsylvania used a significant amount of NG, the NG market was still limited to the western part of the state. The extent of the market was even more limited in New York. Since these states were highly urbanized, contained 18 percent of the housing units in the United States, and could easily be reached by long-distance pipelines from major producing areas, they represented significant potential markets.

The NG production and distribution system within the largely rural states of Wyoming and Montana did not reach many households, but those that were reached used NG extensively. An examination of HDDs statistics indicate that households in these states had very high levels of space heating requirements and many RNG customers met these requirements with NG.

Table 5.1

Statistics for Residential Customers in the United States in 1940

State	Use	Number	Av. Use	Cost	Degree days	Units	Extent	Income
CA	74,795	1,694	44	76.2	2688	2,138	79%	832
OH	65,842	1,221	54	63.9	5906	1,899	64%	654
PA	40,132	692	58	61.2	5895	2,516	27%	646
TX	39,714	649	61	69.3	2036	1,678	39%	429
OK	23,746	257	93	45.3	3564	610	42%	370
WV	21,116	192	110	36.3	5265	445	43%	398
MI	20,406	573	36	103.4	6824	1,396	41%	675
IL	19,269	1,257	16	122.7	6168	2,193	56%	744
KS	18,484	214	36	58.5	4968	511	42%	425
NY	16,934	411	41	80.1	5955	3,662	11%	862
MO	13,624	386	35	84.9	5082	1,069	36%	513
LA	10,656	192	56	69.5	1803	593	32%	359
KY	9,769	172	57	56.6	4549	699	25%	318
AK	6,515	72	91	50.1	3325	496	15%	258
MT	6,178	45	138	46.4	8219	160	28%	565
CO	6,073	98	62	77.8	7106	316	31%	542
MD	5,659	225	25	76.7	4812	466	48%	702
NE	5,532	121	46	75.7	6406	361	33%	436
GA	5,455	86	63	84.5	2797	752	11%	334
MN	5,438	160	34	90.8	8736	728	22%	521
IA	5,034	133	38	99.3	6900	702	19%	498
MS	4,046	49	83	65.5	2519	535	9%	213
TN	3,435	50	69	74.8	3965	715	7%	337
IN	3,346	148	23	113.3	5873	961	15%	547
WY	3,009	22	137	46.6	7992	316	7%	597
UT	2,720	32	85	68.4	6539	140	23%	478
NM	2,196	27	82	69.9	4745	130	21%	375
AL	1,867	33	56	99.4	2840	674	5%	279
SD	1,289	17	76	76.7	7671	165	10%	360
AZ	1,219	37	33	127.9	2301	131	28%	501
FL	148	4	36	133.1	727	520	1%	516
US	443,646	9,245	48	71.1	4694	34,855	27%	589

Source: Use, Customers, and Price, *Minerals Yearbook*, 1940, various pages; Heating Degree Days, National Oceanic and Atmospheric Administration, *Historical Climatology Series, U.S. State, Regional, and National Monthly and Seasonal Heating Degree Days Weighted by Population*, Asheville, N.C., National Climate Center; Occupied Housing Units, Bureau of the Census, *Census of Housing 1940*; Income, Bureau of Economic Analysis, *State Personal Income 1929-1982* (Washington, D.C.: U.S. Government Printing Office, 1984), 8.

Note: Use is consumption in million cubic feet. Number is number of customers in thousands (rounded). Av. use is average use per customer in thousands of cubic feet (rounded). Cost is price per thousand cubic feet. HDDs is heating degree days for the year. Units are occupied housing units in thousands (rounded). Extent is the number of customers relative to the number of occupied housing units. Income is personal income per capita.

Although much of the data in Table 5.1 can be used selectively either to explain why markets developed in certain places or to predict where markets might develop because of weather and economic conditions, the data cannot be used generally to explain differences in the level of use per customer between states because markets had not developed sufficiently in all states to enable such analysis with available data. Not too surprisingly when gas use per customer is regressed on price, income and HDDs, only price is significant at the 5 percent significance level. The correlation coefficient between use per customer and price at -0.73 is, however, quite strong.

During and after the war years, several states would develop into mature RNG markets as the delivery technology improved greatly and as a large proportion of the RNG customers in many states began to use NG for space heating. The fuller development of these markets and the availability of data documenting this growth would facilitate using regression analysis to analyze use per customer.

In particular, the technology of underground storage was necessary for the accelerated growth of RNG markets. In fact, during the early years of accelerated growth in the latter part of the 1940's, the NG industry experienced growing pains because of a lack of adequate underground storage capacity near end-use markets. During the extremely cold winter of 1947-1948 some residential customers were actually evacuated from their homes[3] due to deliverability problems.

TECHNOLOGICAL AND OTHER IMPORTANT CHANGES DURING THE WAR

During the war years the technology of the gas industry was advancing in several important ways that would improve the welfare of consumers.[4] In Cleveland, Ohio, an LNG storage facility holding 150 million cubic feet (MMcf) was completed primarily to serve RNG customers during their peak winter demand periods. This amount of storage could serve 3,409 RNG customers for a year in Ohio, and the capital and operating costs of the plant were considered moderate.[5]

New processes for obtaining ammonium, certain anesthetics, and alcohol, which required NG, were being developed. Crops were being dried with NG. Some of these processes resulted in a qualitative improvement in products for consumers. For example, agricultural products retained their vitamin A content if they were dried with NG instead of the sun, as was typical.

Carbon black was produced from NG gas by much more efficient methods during the war years as the demand for carbon black along with high octane gasoline from NG increased greatly in support of the war effort. Carbon black was used in the production of synthetic rubber products especially tires. As the efficiency of carbon black production improved and the need for synthetic rubber after the war years diminished, demand for NG by carbon black manufacturers declined, and the long-term supply of NG for residential customers increased.

Pipelines for transport of oil extending 1,000 miles were efficiently produced and put in place during the war, with government funds. Several of these pipelines would be used to transport NG after the war.

In 1944, a 1,265 mile pipeline was laid from Corpus Christi, Texas, to Cornwall, Virginia. The first pipe was welded on January 10, 1944, and gas flowed by October 31, 1944. Thus, the pipeline was completed at a rate of 4.3 miles per day.

An increasing number of utilities switched from MG to NG. More than eighty municipalities in fifteen states were supplied with NG for the first time in 1940. However, there were restrictions placed on hookups of new residential space heating customers throughout the war years, and on June 15, 1943, the Office of Price Administration allowed manufacturers to produce and sell cooking and heating stove units after September 1, 1943, only if the material was already in stock or was acquired by authority of the War Production Board.

Appliances such as gas air-conditioning units were developed in readiness for manufacture and sale when the industry would resume peacetime activities. The cost of the air-conditioning units was considered too expensive for widespread initial use, but the cost was expected to decline substantially as experience was gained in producing them. However, equipment and installation costs were considered greater for gas-fired units than for electrical units and electrical units were favored by mass production methods.[6] Since the cost of either NG or MG was still much less than the cost of electrical energy, gaseous fuels would still have a clear competitive advantage over electricity in many markets.

Utilities were very interested in increasing the number of their space heating customers because of the large amount of NG that each space heating customer represented. Thus, gas utilities were actively engaged in selling and servicing such units to encourage this growth.[7] On the other hand, they were very concerned about their ability to fully service this demand because of the great variability in space heating demand between the hours of the day and the months of the year.

Productivity in many industries increased during the war years due to technological improvements. The supply of available labor, however, was also less and this caused wage rates to increase. The increase in wage rates, moreover, would positively influence the growth of RNG after the war. Since the NG industry had a much lower labor cost component than most other industries, this tended to reduce the cost of NG relative to most other goods.

Since NG space-heating units were generally a superior source of space heating energy for the household than coal units, the consequent rise in money income and the relative constancy of the average cost of NG would increase the rate at which NG units were substituted for coal units. The ready availability of consumer credit and growth in household wealth would also increase the demand for NG furnaces and other appliances.

Figure 5.1

The Great Growth in Natural Gas Reserves after 1940

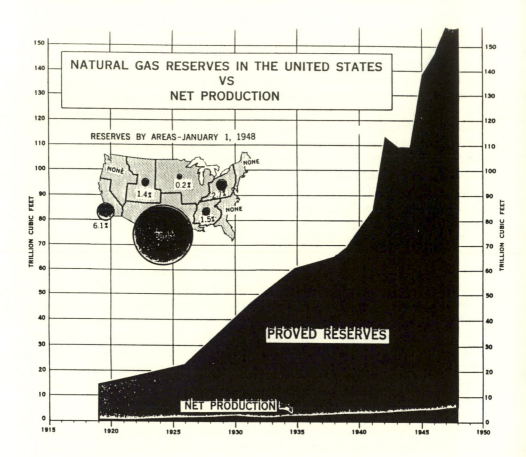

Source: Federal Power Commission, Natural Gas Investigation, Smith-Wimberly, 45

NG use per customer increased systematically after the war. From a 1939 level of 44 Mcf, average NG use per customer in the United States increased in five of the six years between 1939 and 1945 to a level of 55 Mcf in 1945. This increase occurred despite restrictions on NG space heating hookups and on appliance purchases that occurred during the war years.

GROWTH AFTER THE WAR AND TECHNOLOGICAL IMPROVEMENTS

The growth in the use of and the infrastructure development for the relatively more expensive and less convenient liquefied petroleum gas, LP-gas or LPG, was indicative of the failure of the NG market to develop fast enough. In 1940 the amount of LP-gas sold was about 150 million gallons, and by 1949 it was 2,800 million gallons. The majority of this gas was for residential purposes. During the 1940's the first LP-gas tanker and the first large scale LP-gas marine terminal were placed in service.[8] Nonetheless, the NG industry was poised for great growth at the end of the war years. For example, estimated reserves had grown dramatically since the 1930's, reserves were about forty times as great as annual marketed production, and 88 percent of these reserves were in the south-central part of the United States (see figure 5.1), which meant that more pipeline systems would have to be built from the south central to the northern United States to reach production capabilities.

By the end of the 1940's, 30-inch pipelines were being put in place. This was significant because the capacity of a 30-inch pipeline was three-and-one-half times as great as a 16-inch pipeline, the common size for pipelines during the 1930's, and because the increase in cost of a 30-inch line was less than proportionate to the increase in the capacity. The tensile strength of the pipeline of the 1940's was also greater than the tensile strength of the pipeline of the 1930's, which increased the pressure that the pipeline could withstand and, thus, increased further the capacity of the pipeline. The larger and stronger pipelines also offered greater opportunities for using the pipeline system as a storage system at times, thus reducing the uncertainty of supply during the winter periods of peak domestic space heating demands.

By connecting pipeline of different diameter (such as a 24-inch line connected to a 26-inch line, connected to a 30-inch line), the 30-inch line was packed with NG under high pressure which could be withdrawn during periods of high demand by reducing the pressure. This technique became known as line packing.

Great improvements were also made in laying pipeline, such as aerial surveying for selecting the route for new pipelines and pressure welding for joining pipeline sections. These and other improvements in technology reduced the cost at which NG could be supplied to NG customers.

Although steel shortages after the war slowed down some pipeline developments, relatively low interest rates for borrowing funds to build pipeline systems and the greater efficiency in laying pipe resulted in dramatic annual increases in the average number of pipeline miles completed or planned in any one year. As a consequence, many new markets were created. In 1947, the El Paso Natural Gas Company completed a 1,200-mile pipeline system from the Permian Basin in Texas to California, creating a market for Texas NG in California for the first time. High-pressure centrifugal compressors for gas transmission, which had the effect of quickly increasing or decreasing the amount of NG flowing between two points, were also being regularly installed. The 1,840-mile 30-inch pipeline system from South Texas to New York used steam-turbine-drive, high-pressure centrifugal compressors. This pipeline consisted of 1,210 miles of 30" line, 558 miles of 26" line, and 71 miles of 20" line, with 6" to 16" lateral lines and was estimated to have an ultimate capacity of 500 million cubic feet daily, which was equivalent to 3.8 million domestic consumers daily consumption in the United States in 1940.

The 24-inch "big inch" 1,340-mile pipeline system (the world's largest volume oil pipeline system when constructed) and the 20-inch "little inch" 1,475-mile pipeline system (the longest pipeline system in the United States when constructed) were state-of-the-art pipeline systems initially constructed for the transport of oil. They were sold to the Texas Eastern Transmission Corporation for the transport of NG after the war at a cost of $143.1 million, which was less than the original cost of $145.8 million as reported by the War Assets Administration. Centrifugal compressors were used to move large volumes of NG under high pressure over long distances for the first time for the "big inch" and "little inch" lines. Studies indicated that excluding fuel or power the operating cost would be about one-half the operating cost of the commonly used reciprocating compressor equipment.

The Texas Eastern Transmission Corporation commenced operating between Texas and New York in 1947, and it was joined by the Transcontinental Gas Pipeline Company in 1949. In addition to the decrease in the cost of transporting NG, the increase in pipeline construction activity between Texas and northeastern and western markets was encouraged by the high price that could be obtained for NG in these markets, by the complete depreciation of MG equipment, and by newer estimates of the NG resource base in Texas. For example, the Hugoton field in Texas continued to be developed. This field contained estimated reserves of between 22 to 29 trillion cubic feet (Tcf) or 15 percent of the known United States reserves.

The Need for Balanced Growth

The major technological hurdle during the 1940's for further improvement in the NG delivery system was not the technology of pipeline construction but was still NG storage or, more generally, techniques for reducing or better satisfying variations in the level of domestic demand for NG service during the day and over the course of a year. The NG industry used as an indicator of the economic significance of this variation in demand the ratio of average monthly or daily demand to the maximum monthly or daily demand. This indicator was referred to as the load factor. The smaller the load factor, the greater the cost of serving a customer. The Bureau of Mines in its Technology Section of the 1949 Minerals Yearbook stated that most technical developments within the industry were aimed at reducing or eliminating load problems. It also cited developments that were addressing the problem. "The Federal Power Commission has granted one company in the Chicago area permission to construct a plant to provide and store liquid natural gas. It will have a storage capacity of 400,000 million cubic feet. . . . In the appliance industry the peak load problem is being attacked by the introduction of gas-oil furnaces that would burn gas under normal conditions but could switch to oil during high demand periods."[9]

In order to address the load problem, a few gas utilities, especially MG utilities, were also studying much more precisely the variability in RNG demand. Utilities were studying the incremental demand from the addition of particular appliances. For an average MG customer of a utility in Philadelphia it was estimated that the addition of a central space heating unit for a customer who used gas for cooking and water heating increased gas use five times.

Some utilities were randomly selecting households from their customer lists and placing special instruments in these households to measure gas consumption on an hourly basis for several months. Detailed graphical displays of the relationship between gas use and temperature were prepared. From these graphs it was determined that demand reached a peak between 6:30 A.M. and 7:30 A.M.; that the relationship between temperature and use was nonlinear between 6:30 A.M. and 10:00 A.M.; and that the relationship was linear for the remainder of the day. Demand declined by about 30 percent between 7:00 A.M. and 10:00 A.M., then remained relatively constant until 9:00 P.M. and then declined until about 5:00 A.M.. It was also discovered that demand in January could be ten times as great as demand in August.[10,11] Accordingly, load factors were quite low.

In order to increase summertime demand for NG and thus increase load factors, the gas industry was promoting gas air conditioning, and between 200 and 800 units were sold in 1949 in some southern towns and even in some northern towns.[12] Despite some initial successes, air conditioners were never widely enough used to greatly affect load factors.

Growth in Underground Storage

With lackluster growth in the use of NG air conditioners and refrigerators, with an increased demand for space heating service and for reliable NG service, the need for additional underground storage facilities was clear. Accordingly, the amount of NG stored underground grew from 36,167 million cubic feet in 1945 to 175,260 million cubic feet in 1950.

A comparison of RNG consumption, NG production, and NG underground storage facility statistics for Pennsylvania, Ohio, and West Virginia in 1910 and in 1949 highlights the importance of underground NG storage to RNG markets. In 1910 these three states accounted for 69 percent of total United States RNG consumption, 72 percent of total United States NG production, and they contained no underground NG storage facilities. In 1949, they still accounted for a sizable proportion (25 percent) of total United States RNG consumption, but they only accounted for 5.8 percent of total United states marketed production. Yet, they withdrew 53 percent of the total amount of NG withdrawn from storage sites in the United States in that year.

Relatively inexpensive NG was being transported regularly from the Southwest and stored in depleted oil and NG reservoirs during the summer months in Ohio, Pennsylvania, and West Virginia. It was then withdrawn during the winter months to satisfy residential space heating demands. This process had partially replaced the production of NG from wells in the vicinity of consumption in the Appalachian region. If underground NG storage reservoirs had not been developed, huge pipelines would have had to be built between the south-central and the northern parts of the United States in order to adequately satisfy the increased winter peak demands for NG space heating. In the winter these pipelines would be operated at near capacity. In the summer, without the extensive use of NG air conditioning, they would be operated at between 10 percent and 25 percent of capacity. The economic cost of such a system would have reduced greatly the economic advantage of NG as a household fuel.

INTERACTION BETWEEN NATURAL GAS SERVICE, SOCIAL BEHAVIOR, AND MARKET GROWTH

In addition to reducing seasonal load problems, the extensive use of NG in new appliances was perceived by some as altering the way in which houses were constructed. Air conditioning would eliminate the need for porches and basements and would reduce the number of outside doors and windows. Air conditioning, along with dryers and incinerators, would enable homemakers to be almost completely independent of the weather. The homemaker needn't go outdoors on an unpleasant day to take out the trash or to hang out the clothes but could perform these tasks indoors with appliances.[13] Just as increased application of NG and other energy equipment was affecting individual behavior, changes in individual behavior were affecting the demand for appliances and the type of energy used in the household.

Americans were marrying younger. The median age of females at the time of their first marriage was more than a full year less in 1950 than it was in 1940 (see

Table 5.2). The age of men at the time of their first marriage had also declined, and both the age of men and the age of women at the time of their first marriage were near their lowest levels ever. Accordingly, Americans were having more babies. Birth rates in the 1940's reached their highest levels since the 1920's. The birth rate of 1947 at 26.6 births per thousand population has not been surpassed since. The growth in family formation, especially the birth of babies, increased the need for heating, cooking, bathing, and cleaning services within the household. These needs could be most efficiently satisfied by appliances, furnaces, and boilers that used gas and oil rather than coal. The purchase of NG service for space heating, water heating, and cooking could eliminate certain unpleasant activities required in obtaining these services from coal and wood burning equipment, such as cleaning up the area after supplying the equipment with fuel and taking away the ashes and soot after combustion of the fuel. Starting a new fire in the furnace or stove when it unexpectedly burned out could be a particularly time consuming and annoying chore.

Table 5.2

Demographic and Economic Factors That Influenced Market Growth

	Age at 1st marriage		Birth rate	Unemployment rate	Recreation investment	Female participation
	Male	Female				
1910	25.1	21.6	30.1	5.9		
1920	24.6	21.2	27.7	5.2		20.4
1930	24.3	21.3	21.3	8.7	921	21.9
1940	24.3	21.5	19.4	14.6	494	24.6
1950	22.8	20.3	24.1	5.3	2421	28.8
1960	22.8	20.3	23.7	5.5	3412	32.2
1970	23.2	20.8	18.4	4.9	8328	36.7

Note: All reported statistics from *Historical Statistics, Census.* The page number where the statistic(s) can be found in this publication is reported after the designated series. Age at 1st Marriage is Median Age at First Marriage, 49. Birth rate is expressed as total births per thousand population, 49. Unemployment rate is expressed as a percent of civilian labor force, 435. Recreation investment is Household Recreation Investment, i.e. radio, television receivers, records, and musical instruments expressed in millions dollars, 401. Female participation is females as a percentage of total labor force, 136.

The unemployment level was also much lower during the 1940's than it was during the previous decade. In 1948 only 3.8 percent of the civilian labor force was unemployed. Thus, the amount of manpower available around the household for maintaining coal and wood furnaces, boilers, and ranges was reduced, and the opportunity cost of continuing such activities was higher.

Capital expenditures on televisions, record players, and other sources of household entertainment had increased greatly after the war years. Gas service could enable the investor in these products to allocate more time for the appreciation of the services that they had to offer. The percentage increase in expenditures for this category of household expenditure (see Table 5.2) was one of the largest percentage increases in expenditure for any item between 1945 and 1950. Other growth markets were furniture equipment and supplies, new and used cars, and foreign travel.[14]

Although the percentage of females in the total labor force after the war was less than it was during the war, this percentage was still much higher than it was during the previous decade. Women were not only increasing their productive activity in the household - by increasing the number of appliances they used and the number of babies they were having - they were also still maintaining relatively high levels of employment outside the household, which, in turn, increased average household income and the demand for and the ability to pay for new NG appliances to replace inferior ones.

Table 5.3

Furnace and Appliance Stock Purchases after the War (in 1000's of units)

| | Furnaces | Water heaters | | Ranges | |
	Gas	Gas	LPG	Gas	LPG
1945	52				
1946	230	1,189	139	1,440	360
1947	188	1,430	370	1,816	618
1948	188	1,316	184	2,075	675
1949	299	1,283	183	1,608	499
1950	600	2,054	310	2,396	635
Total	1,557	7,272	1,186	9,335	2,787

Sources: Gas-fired Warm Air Furnaces, Gas Appliance Manufacturers Association; Heater and Range Sales, American Gas Association.

Note: Furnaces are Gas-fired Warm Air furnaces. Water Heaters are Domestic Water Heater Sales and gas is city gas. The numbers were estimated by the American Gas Association from Gas Appliance Manufacturers data in which shipments by manufacturers were expanded from returns of reporting companies to represent industry-wide coverage. Shipments of appliances for LP-gas users include those obtaining such gas through utility mains, as well as bottled gas uses.

Some of the increase in the demand for gas service was reflected in the number of gas appliances sold after the war years (see Table 5.3). Since there were shortages of materials for producing appliances after the war years, and NG service was not generally available in many places where it was demanded, these statistics understate the change in demand for NG service. Nonetheless, the statistics do indicate that the demand for NG, based on the demand for appliances, had increased greatly from 1930 levels.

The market for appliances, however, was not free of problems. During the 1930's and 1940's, the cost of servicing gas appliances was increasing for many utilities as the number of automatic gas appliances increased.[15] Many gas utilities were sellers of gas appliances as well as sellers of gas, and they had an interest in retaining customers by keeping customers, who had become increasingly dependent on them, contented. Because state utility commissions allowed them to pass the cost of servicing particular customers on to all customers, many gas utilities provided free service for gas appliances. For some gas maintenance men, the changes that had taken place took on an ominous air because in the past there was always an old coal stove or gas range that could substitute for an inoperable gas stove or range. However, customers no longer kept standby equipment. "If the range fails they are seriously inconvenienced. If the refrigerator fails food spoilage is almost certain; and when we advocate house heating by gas, we must do so fully realizing that lack of service on our part may make uninhabitable the homes of our customers."[16]

TYPES OF CUSTOMERS AND PRICES

Mixed gas was MG which also contained some NG. Even if the gas contained a very small proportion of MG, it was still designated "mixed gas" and not NG. Mixed gas was generally less expensive than MG but more expensive than NG.

Table 5.4

Residential Gas Customers by Type of Gas Used (in 1000's)

	Natural	Manufactured	Mixed	LPG	Total
1945	8,212	8,053	2,269	72	18,606
1946	8,594	8,171	2,281	110	19,156
1947	9,337	8,073	2,216	210	19,836
1948	10,492	7,933	1,849	288	20,562
1949	11,409	7,877	1,656	322	21,264
1950	13,083	7,080	1,684	298	22,145
1955	21,085	1,160	3,819	219	26,283
1960	28,087	164	2,054	113	30,418
1965	33,591	119	579	51	34,340

Source: American Gas Association, *Historical Statistics*, 1965, Table 79.
Note: LPG is liquefied petroleum gas.

Accordingly, the number of mixed gas customers declined at a slower rate than the number of MG customers, as NG became increasingly available (see Table 5.4). As the equipment used to produce mixed and MG was fully depreciated, as contracts to buy MG from non utilities expired, and as the number of underground NG storage facilities increased, the number of MG customers and mixed gas customers would decline.

In most places NG service was generally less expensive than MG service. There were exceptions. In some parts of the Northeast that were far from ready supplies of NG and where utilities had access to relatively inexpensive coke-oven gas or had particularly efficient gas manufacturing plants, the price of MG was not much different from the price of NG. Nonetheless, NG resulted in less environmental damage than MG, and NG was non toxic, whereas the most common MG, carburated water gas, contained significant amounts of toxic carbon monoxide. For such reasons NG was considered by many to be the superior fuel even when the price of both fuels was similar.

The number of LPG customers continued to increase for several years after the war from increased demand for gas of all kinds relative to other fuels. However, the number of LPG customers began to decline as NG became more readily available.

Table 5.5

New Residential Customers and Related Economic Statistics

	New customers[a]	New units[b]	Capital formation[c]	Wealth[d]
1945	291	208	$752	$78
1946	512	662	$2,479	$79
1947	732	846	$3,270	$81
1948	1304	914	$4,002	$83
1949	1182	989	$3,956	$85
1950	2216	1352	$5,763	$89

Sources: New Dwelling Units, Capital Formation and Wealth, Grebler, Blank, and Winnick, *Capital Formation in Residential Real Estate*, 332, 338, 361; New Customers, *Minerals Yearbook*, various years.
  [a] Change in the number of residential customers between years in thousands.
  [b] New private permanent nonfarm housekeeping dwelling units started in thousands.
  [c] Gross capital formation in housekeeping residential real estate (1929 dollars in millions).
  [d] Cumulated estimates of nonfarm residential wealth - structures (1929 dollars in billions).

After having declined during much of the 1930's, gross capital formation in residential real estate and nonfarm wealth increased dramatically after the war. This change, along with income growth and other changes discussed previously, supported further growth in the number of RNG customers. Accordingly, the magnitude of the change in the number of customers between 1945 and 1950 was greater than the change in the number of new housing units started (see Table 5.5). Moreover, the percentage change in the annual number of new customers between 1945 and 1950 exceeded the percentage change in wealth and capital formation.

Between 1945 and 1950, the rate of growth in the revenue received from RNG customers also exceeded the rate of growth in the number of gas customers. This occurred because an increasing proportion of RNG customers were receiving space-heating service. The percentage change in revenues received from RNG customers between 1945 and 1950 increased greatly while the nominal cost of NG was relatively constant (see Table 5.6). The average deflated cost of NG would not rise to 1945 levels until 1960.[17]

The deflated price of NG fell by 25 percent between 1945 and 1949. This means that if all customers who were gas customers in 1945 continued to be willing to pay the 1945 price for gas service, and if this price decline had not entered into the decision to purchase gas service, then these customers experienced an unexpected revenue stream, or windfall, totalling $88 million, when expressed in constant dollars.[18]

Table 5.6

Residential Gas Revenues, Sales, and Price between 1945 and 1965

Year	Revenues[a]		Demand[b]		Nominal Cost[c]		Real Cost[d]	
	NG	MG	NG	MG	NG	MG	NG	MG
1945	353	269	5,601	1,365	.63	1.97	.63	1.97
1946	379	287	6,163	1,466	.62	1.96	.57	1.81
1947	455	307	7,514	1,558	.61	1.97	.49	1.59
1948	529	341	8,728	1,584	.61	2.15	.46	1.61
1949	591	355	9,541	1,526	.62	2.33	.47	1.76
1950	749	454	11,562	1,454	.65	3.12	.49	2.33
1955	1,657	59	20,086	235	.82	2.49	.55	1.67
1960	2,939	15	30,231	64	.97	2.29	.65	1.54
1965	3,938	13	39,190	63	1.00	2.12	.57	1.21

Sources: Revenue, Demand, Price, American Gas Association, *Historical Statistics of the Gas Industry, 1965*, Tables 95, 97, 115, and 117; Deflator, CPIU-all items, *Historical Statistics, Census*.

[a] In millions of dollars.

[b] In millions of therms.

[c] Price per 10 therms.

[d] Price per 10 therms in constant dollars.

The cost of coal increased consistently throughout the 1940's. Yet, its cost was still less than the cost of NG on a Btu equivalent basis in those places where both fuels were generally available. Moreover, by 1950 articles had appeared in such financial papers as Barron's in which the cost of heating a home with NG was generally considered less expensive than heating a home with coal.[19]

The price of fuel oil, which had also increased systematically along with the price of coal during the 1940's, rose to levels where it was clearly much more expensive than NG when prices were adjusted to reflect the differential efficiency at which these fuels were burned in space-heating units (see Table 5.7). Even when available energy prices were not adjusted for the greater efficiency at which space heat was obtainable from a NG furnace, fuel oil prices were above NG prices in many markets.[20]

NG was less expensive than the generally used alternative sources of space-heating energy in such major cities as Atlanta, Chicago, Cincinnati, Cleveland, Detroit, Kansas City, Minneapolis, Pittsburg, and St. Louis (see Table 5.7). NG was the dominant fuel and very inexpensive in San Francisco and Houston. However, in such major northeastern cities as Philadelphia and New York, gas was more expensive than fuel oil in 1949. Much of this gas was still MG. Gas markets in these eastern cities had developed much more slowly than gas markets in most other parts of the United States.

NATURAL GAS MARKETS AT THE CLOSE OF THE 1940's

By 1950 RNG markets had evolved to occupy a position both within the NG industry and within the economy that was much different from earlier decades. Four distinct developments were the growth of sizable markets within and between many states, the growth in the type and number of automatic NG appliances sold, the allocation of increasing amounts of NG from carbon black and field use markets to residential markets, and improvements in NG conservation.

By 1950 sizable markets had developed within and between many states similar to the earlier development of an interstate market in West Virginia, eastern Ohio, western Pennsylvania, and southwestern New York during the first decade of the twentieth century. The new interstate markets of 1950, however, generally extended over a much wider area within each state.

Significant markets had developed even in relatively unpopulated parts of the country such as the northwestern regional market in Wyoming, Montana, and Colorado, where a large proportion of the households used NG extensively, and where there was much trade in NG between states. For example, the Montana Dakota Utilities Company began transporting NG on a 340-mile line between Montana and Wyoming in 1950, which resulted in formerly wasted NG in Wyoming being shipped to Montana.

Table 5.7

Cost of Residential Heating Fuels by City (in cents per million Btu)

	Type of Fuel	1941	1945	1946	1947	1948	1949
Atlanta	Bit. Coal	67	74	90	107	119	119
	Gas	79	77	75	69	64	64
Chicago	Bit. Coal	70	78	86	107	121	123
	Fuel Oil	NA	NA	NA	NA	NA	166
	Gas	88	88	88	88	88	88
Cincinnati	Bit. Coal	63	70	78	96	109	109
	Gas	70	65	65	73	73	73
Cleveland	Bit. Coal	73	81	90	111	124	126
	Gas	63	63	63	63	64	64
Detroit	Bit. Coal	73	82	89	113	120	125
	Fuel Oil	NA	NA	NA	NA	NA	176
	Gas	82	78	78	77	77	87
Houston	Gas	73	67	68	68	67	68
Kansas City	Bit. Coal	38	43	42	60	67	65
	Fuel Oil	94	99	116	167	172	154
	Gas	61	61	61	55	56	57
Los Angeles	Gas	58	58	58	58	58	58
Minneapolis	Bit. Coal	99	114	124	151	172	175
	Fuel Oil	95	103	117	173	179	147
	Gas	96	80	78	79	79	83
New York	Anth. Coal	71	85	96	NA	NA	126
	Fuel Oil	92	96	115	153	161	150
	Gas	121	121	121	131	131	186
Philadelphia	Anth. Coal	69	86	96	105	116	118
	Fuel Oil	91	94	109	140	152	147
	Gas	118	118	118	142	156	156
Pittsburgh	Bit. Coal	38	40	47	75	83	78
	Gas	60	60	60	61	61	61
Portland	Bit. Coal	128	135	141	166	176	178
	Fuel Oil	76	77	105	122	147	147
	Gas	89	89	108	175	206	159
St. Louis	Bit. Coal	46	52	60	78	87	87
	Gas	77	77	77	77	77	76
San Francisco	Gas	53	48	42	42	42	42
Scranton	Anth. Coal	49	64	74	78	87	87
	Gas	150	150	150	150	150	150
Seattle	Bit. Coal	114	139	146	164	173	177
	Fuel Oil	82	82	113	131	160	160
	Gas	114	114	126	126	190	189
Washington	Anth. Coal	73	86	96	106	116	121
	Fuel Oil	96	100	118	138	156	152
	Gas	127	130	130	130	130	132

Source: American Gas Association, Historical Statistics of the Gas Industry, Table 244.

Note: Prices are rounded and represent fuel costs with differential efficiencies for the heating equipment by fuel: gas 80 per cent; anthracite 62 per cent; bituminous coal (stoker) 59 per cent; oil 57 percent; bituminous coal (hand fired) 48 per cent. NA indicates data not available.

The significant east-north-central market between the states of West Virginia and Ohio had continued to grow once more as the number of NG customers, and in particular the number of space-heating customers, continued to increase.

A significant central regional market had developed in Minnesota, Nebraska, Iowa, Missouri, and Kansas.

The pipeline system from Texas through New Mexico and Arizona had evolved into a southwestern market along the way to California. This pipeline system had contributed to the further development of California as a market which by 1950 had 75 percent more customers than Ohio, which had the next largest number of customers.

An extensive market had continued to develop within and between Texas, Louisiania, and Oklahoma, which had become by far the major producing area and the major industrial market for NG in the United States. In 1950 Texas alone produced 50 percent of the NG produced in the United States, whereas it had produced only 1 percent in 1911. Industrial establishments in Texas also consumed about 40 percent of the NG consumed by such establishments in the United States. Texas had in fact developed an industrial policy in which state representatives attempted to attract firms to the state with the lure of cheap energy, especially NG.

There were seventeen states where significant markets had developed by 1950 and where the number of gas customers as a percentage of the population exceeded 10 percent (see Table 5.8).[21] For example, Ohio was a very significant market. In Ohio a large proportion of the large number of households in the state received NG service; a significant proportion of these customers were space-heating customers; and the winters were cold. The level of use for the other states also varied depending on the proportion of space-heating customers to the total number of customers and on the level of heating degree days.

The number of customers and the proportion of space-heating customers to total NG customers would continue to grow during the 1950's within the states listed in Table 5.8 except for the states of Wyoming and Montana. By 1945 almost all customers in Wyoming and Montana appear to have been space heating customers. After 1945 only the number of customers would increase; the proportion of space-heating customers to total domestic customers would remain near one. Since the proportion of space-heating customers to total NG customers was near one in all states listed in Table 5.8 by 1960, the differences in sales per customer between states in this year were largely a function of differences in weather and in price.

Data regarding RNG consumption in states such as New York and Pennsylvania, in which the aggregate amount of NG consumed was still large relative to the aggregate amounts consumed in other states, are not included in Table 5.8 because NG was still not generally available throughout the state in 1950 and the proportion of NG customers to the total population of the state was small. However, this situation would change during the 1950's.

Table 5.8

Residential Customers and Sales per Customer in Major Gas Using States

	Customers				Sales per Customer			
	1945	1950	1955	1960	1945	1950	1955	1960
WV	205	255	299	325	118	147	158	164
OH	1165	1501	1762	2039	75	124	168	183
MI	815	975	1209	1372	40	81	103	147
MN	33	275	336	433	79	92	140	149
IA	100	202	323	389	47	82	120	152
MO	193	541	655	756	76	95	125	150
NB	76	163	217	252	83	113	140	161
KS	249	342	430	483	104	141	149	157
OK	291	387	480	541	103	105	108	115
TX	811	1217	1678	2021	65	68	76	87
LA	236	364	502	645	57	64	75	87
MT	49	66	82	105	169	187	187	176
WY	26	38	49	57	175	167	189	174
CO	114	164	262	342	70	131	147	150
NM	37	74	120	157	95	99	114	134
AZ	47	106	189	271	53	52	68	75
CA	1962	2605	3406	4172	64	74	92	93

Source: American Gas Association, *Historical Statistics of the Gas Industry, 1965*, various tables.

CONCLUDING COMMENTS

The NG market after the war years had changed in ways that both directly and indirectly improved the economic welfare of households. The changes were also, interestingly enough, consistent with the types of changes that members of the Cities Alliance would have liked to have brought about by legislative means during the 1930's.

In 1948 the amount of RNG consumption exceeded for the first time the amount of NG that was vented or flared and wasted in the production of NG and oil. The amount vented and flared would continue to decline in later years, and by 1965 it was less than 10 percent of the amount of NG used by residential customers. This decline was the type of conservation initiative that had been deemed desirable by national and state agencies representing the public interest since the nineteenth century.

The production of carbon black, produced ordinarily from sour (high sulfur) NG, released to the environment huge amounts of sulfur, carbon and particulates damaging to the ecosystem. Moreover, the heat value of NG was wasted when carbon black was produced. Nonetheless, the development of this industry was not controlled by public bodies. Instead, the industry was allowed to develop whenever and wherever short-run profits could be gained. Clearly, many representatives of the Cities Alliance had thought, but had not fully articulated, that the suitably discounted future value of the heat value of NG to residential markets, and the cost of the gas distribution system required to deliver NG to these markets, was not being considered in the allocation of NG to markets. Thus, the government could have interceded for the public interest, as it had interceded when it had attempted to make electricity available as cheaply as possible to a broad segment of the public, hence the public power programs and rural electrification. Instead, it was only after the war years, rather than the 1930's, that the amount of NG used to produce carbon black began to decline (see Table 5.9) and the amount of NG used in households began to rise dramatically.

Table 5.9

Significant Changes in Natural Gas Use (in billion cubic feet)

					Column Entry Ratios		
	Residential	Vented/flared	Carbon black	Field use	2/1	3/1	4/1
1940	444	656	369	712	1.48	0.83	1.60
1945	607	896	432	917	1.48	0.71	1.51
1946	661	1,102	478	898	1.67	0.72	1.36
1947	802	1,068	485	934	1.33	0.60	1.10
1948	896	810	481	1,022	0.90	0.53	1.14
1949	993	854	428	1,060	0.86	0.43	1.07
1950	1,198	801	411	1,187	0.68	0.34	0.99
1955	2,124	774	245	1,508	0.36	0.11	0.71
1960	3,103	563	198	1,780	0.18	0.06	0.57
1965	3,903	319	115	1,910	0.08	0.03	0.49

Source: *Historical Statistics, Census*, 595.
Note: Column entry ratios are the ratio of other uses to residential use. For example column entry ratio 2/1 is the ratio of vented/flared use to residential use.

By 1950, moreover, the use of NG in the production of oil and natural gas from wells or for mining activity at the lease site was also less than residential consumption for the first time. RNG markets had, in a sense, come of age.

Yet some might argue that the industry, constrained in part by regulatory conditions, had made decisions during its coming of age that were not best from the point of view of the residential customer.

During the 1930's and 1940's the gas industry had at least two solutions for improving the economic attractiveness and reliability of NG service. One solution was to fully promote the development of NG air conditioners and refrigerators. Another solution was to increase underground storage investments.

Since new investment in underground storage represents capital expenditures of gas businesses these investments were included in the rate base of gas businesses. Moreover, the amount of income gas businesses were allowed to earn was also increased by these investments since this income was calculated as the product of the rate base and a regulated rate of return on investment. Now if the gas industry had invested significant amounts of money in the further development of NG air conditioning and refrigeration technologies, these investment dollars would have also been included in the rate base and the amount of income gas businesses were allowed to earn would also have increased accordingly. But when gas air conditioners and refrigerators were purchased by RNG customers, these investments would, of course, not be included in the rate base of gas businesses. Moreover, the cost of serving residential customers would also have decreased because of these investments since the pipeline and distribution systems from the wellhead to the consumer would be more intensively used throughout the year. Hence, the load factor would have been increased. Eventually, the reduction in the cost of supplying NG service and the increased purchase of gas air conditioners and refrigerators would have increased significantly the amount of NG purchased by residential customers. The income received by the gas industry would have been increased by these expenditures. But when this occurred the industry would have had to return some of this increase in income to residential customers in the form of rebates since their rate base would not have increased significantly in the interim. Firms in the industry would be receiving a rate of return which was greater than the regulated rate of return. Therefore, the gas industry was less apt to promote and explore investments in air conditioning and refrigeration but more prone to increase investment in underground NG storage.

NOTES

1. S. Lebergott, *The Americans, an Economic Record* (New York: W. W. Norton & Company, 1984), 498.

2. Rose, "Urban Environments," 503-539.

3. Sanders, *The Regulation of Natural Gas,* 61.

4. *Minerals Yearbook*, 1940, 1036.

5. The comments on technology and other information that relates to industy developments are based on information reported in the NG chapters of the Minerals Yearbook for the years 1939-1951 and several other principle sources: *American Gas*

Journal, Diary of an Industry (Dallas, Texas: American Gas Journal Publishing Company, October, 1959); Leeston, Jacobs, and Crichton, *The Dynamic Natural Gas Industry*; *Gas Age* (Duluth, Minnesota: Moore Publishing Company, Inc., various years); J. R. Stockton, R. C. Henshaw and R. W. Graves, *Economics of Natural Gas in Texas* (Austin, Texas: Bureau of Business Research, The University of Texas, 1952).

6. Research in Gas Summer Air Conditioning, *American Gas Association, Proceedings 1941*, 421.

7. H. H. Cuthrell, "Analysis of Survey Conducted for Metropolitan Gas Heating and Air Conditioning Council," *American Gas Association, Proceedings 1941*, 262-296.

8. LP-gas includes mostly propane, ethane, butane, and mixtures of these hydrocarbon gases that are produced when the NG is sent to NG processing plants after being withdrawn from the well. LP-gas was used mostly in rural parts of a state where the convenience of a gaseous fuel was appreciated but where the density of the population was not great enough to justify the extension of distribution lines to these households.

9. *Minerals Yearbook*, 1949, 816-817.

10. C. Bary, "Characteristics of Gas House Heating Loads," *American Gas Association, Proceedings 1941*, 90.

11. Similar results were obtained for a utility in Philadelphia which had 8,000 customers using gas for house heating and another company identified only as "Company A" which had 15,500 customers in a metropolitan area using gas for house heating. The data and details about the sampling and measurement procedure were reported in several articles in the *American Gas Association, Proceedings 1949*, 96-111.

12. H. Massey, "Selling the New Big Jobs," *American Gas Association, Proceedings 1949*, 285.

13. Massey, "Selling the New Big Jobs," 281-285.

14. *Historical Statistics, Census*, 318.

15. W. M. Little, "Technical Aspects of Appliance Servicing," *American Gas Association, Proceedings 1941*, 463-465. W. C. Peters and G. B. Johnson, "Simplification of Gas Appliances and Standardization of Service Manual Material," *American Gas Association, Proceedings 1949*, 462-465.

16. No author, "Sales and Service Relations," *American Gas Association, Proceedings 1941*, 259.

17. The rise in price occurred as increasing amounts of relatively expensive NG were being supplied to the Northeast for the first time between 1950 and 1960. The reasons frequently given as to why this NG was relatively more expensive were: it had to be transported greater distances; pipelines had to be extended into highly urbanized areas, and extensions into highly urbanized areas are more expensive than the average cost of extensions; and the cost of NG at the well-head was increasing.

18. In this calculation it is assumed that price had no effect on gas use per customer. Thus, the calculated value would be higher if price were assumed to have had an effect on gas use per customer. It has been estimated in previous chapters and it will be estimated in later chapters that the own-price elasticity for NG demand per customer is not equal to zero and is probably nearer to -0.5 in value than it is to zero. Thus, the absolute value of the reported number is viewed as a lower bound estimate.

During the late 1930's and 1940's, studies by economists began to appear which considered the quantitative relationship between the price of NG and the demand for NG by customer class. Although relationships were not estimated using regression analysis and a clear distinction was not made between estimating the effect of price on use per customer and the number of new customers in these articles, relationships were presented graphically. The own-price elasticity for residential customers was considered negative and greater than zero but the absolute value of the own-price elasticity for industrial customers was considered much larger than the absolute value of the own-price elasticity for residential customers. The authoritative discussion for the time period is contained in a text by Troxel. Troxel, *Economics of Public Utilitites*, 549-644.

19. J. R. Stockton, "Competition among Fuels", in *Economics of Natural Gas in Texas*, ed. Henshaw and Graves, 247-290.

20. This conclusion is reached by converting average fuel prices reported by the U.S. Bureau of Labor Statistics, *Locally Important Fuels*, 1949, and expressing these prices in equivalent energy units using the conversion formulae contained in Federal Power Commission, *Natural Gas Investigation*, 1948, 329-354.

21. Sales-per-customer amounts reported in Table 40 can be compared with amounts reported previously by using the conversion factor of 10 therms equals a 1,000 cubic feet. Sales-per-customer by 1945 had increased appreciably from the average level of the 1930's for such states as Ohio, West Virginia, and Oklahoma. This increase is consistent with the increase in the proportion of NG customers that were space-heating customers in 1945 and the much more favorable economic conditions to include the reduction in the real price of NG during the 1940's.

6

GROWTH IN THE MARKET
BETWEEN 1950 AND 1973

By 1950 the mature technology of the NG industry was able to expeditiously compress a room full of NG into a space the size of a suitcase, ship the NG within long distance "trunk" pipelines to the Northeast, and then decompress the NG once more to room size in preparation for delivery into a Northeast metropolitan NG market.[1] If more or less NG was required to serve customers within this market, the pressure was increased or reduced accordingly. Centrifugal compressors improved the speed and efficiency at which the pressure was varied to control the amount of NG delivered to markets. This automatic flexibility of supply to demand conditions was unique to NG as a source of energy that was being exploited with greater frequency and with greater success during the 1950's.

The chance of a loss of service to domestic and other customers over a wide area during the wintertime was much less in the 1950's than during the closing years of the Second World War when according to the Petroleum Administration for War (PAW)[2] "300,000 tons of steel for the war program were lost" because of curtailments in gas service. In the 1950's a loss of service could occur primarily for two reasons: either existing pipeline capacity was not great enough to satisfy a sharp sustained increase in demand from a cold spell, or wells froze, and as a consequence overall supplies were reduced.

The Ohio cities of Columbus, Cleveland, and Cincinnati almost lost gas service during the Second World War. If any of these cities had lost service, it would have been out of service for "6 weeks to 2 months" according to the PAW. The potential effect of such a loss of service on the war effort and human life was considered to have been much greater than the loss in steel production because of industrial curtailments. Residential customers were given first priority. NG was diverted from industrial customers to residential customers who were several hundred miles apart. Only the coordinated effort of government and industry made possible the avoidance of a catastrophe.

During the 1950's the industry further demonstrated its maturity by delivering NG for the first time to the far Northeast (New Hampshire), far Northwest (Washington), and to the Carolinas, and by uninterruptedly delivering this NG to all parts of the United States under a wide variety of demand conditions. By the end of the decade, NG was being delivered to domestic customers in every state except Maine, Vermont, and Hawaii.

The regional nature of the NG industry had become increasingly important and social dilemmas created by regional differences had become increasingly complicated as a consequence of the great growth in NG production capacity in the south-central United States. Institutional changes in tax and in regulatory policies were apt to simultaneously improve and reduce the economic welfare of consumers in different parts of the country. For example, a well-head tax on NG in Texas increased the price of NG in Texas and in many states that imported NG. Yet, since a significant portion of these tax dollars was allocated for educational programs in Texas and were used to develop departments with national reputations at Texas universities (which "stole" faculties from northern universities), they worked to provide an inexpensive, but high quality college education for many Texas citizens.

Large reserves, engineering skill, a motivated labor force, and availability of financial resources at reasonable rates supported the overall growth in the 1950's. Shortages in material for building pipelines at the beginning of the 1950's, and bureaucratic logjams at the Federal Power Commission slowed progress at times, but did not seriously impede it.

The financial integrity of the industry enabled an increasing proportion of the moneys required by the industry for expansion to be raised from outside sources.[3] The low rates at which the industry was able to obtain funding tended to keep the cost of NG to domestic customers down. There were technological improvements as well.[4] Pressure gas welding, protective wrapping of pipes with a double coat/double wrap machine, x-ray machines for testing pipeline welds, 36-inch pipes, and high pressure centrifugal compressors were increasingly used during the 1950's. Technological improvements in gas deliverability, low interest rates and the regular review of factors influencing the cost of service by public service commissions tended to keep the cost of gas to residential customers down. According to some the pro-consumer attitudes of the members of the Federal Power Commission were also influencing the price paid for natural gas by residential customers.[5]

The RNG market was a growing market. This growth was supported not only by reductions in the price charged domestic customers but also by continued improvements in the reliability of NG service and by the commonly held view that huge reserves of NG were available to serve current markets, with even more reserves soon to be discovered.

A SUMMARY OF THE GREAT GROWTH IN NATURAL GAS MARKETS BETWEEN 1949-1973

As the 1950's came to a close, the amount of NG used exceeded the amount of coal used in the American economy for the first time (see Table 6.1). This was indicative of the great growth in RNG markets after the war years. As consumers continued to substitute gas furnaces for coal furnaces in the North, the overall importance of coal to the American economy declined.

Table 6.1

Energy Consumption by Type of Energy (in quadrillion Btu)

Year	Coal	Gas	Oil	Other	Nuclear	Total
1950	12.35	5.97	13.32	1.45	0	33.08
1960	9.82	12.39	19.92	1.66	.01	43.80
1970	12.26	21.79	29.52	2.61	.24	66.43

Source: Energy Information Administration, U.S. Department of Energy, *Annual Energy Review 1987*, DOE/EIA-0384 (Washington, D.C.: Energy Information Administration, 1988), 11.

Note: Gas is natural gas. Oil is petroleum. Other is hydropower, geothermal and other sources of energy. Nuclear is nuclear electric power.

The amount of NG relative to the amount of petroleum used in the economy also increased from 43 percent in 1949 to a peak of 74 percent in 1970. Only the enormous fleet of gasoline powered cars, trucks, and airplanes required to support the highly mobile social and economic activity characteristic of modern American society kept NG behind petroleum as the primary energy source in American society during this great growth period.

The rate of growth in the consumption of NG by all major sectors between 1949 and 1972 (see Table 6.2) exceeded the rate of growth in real GNP, number of housing units, and other common measures of economic growth in the American economy by several orders of magnitude. The great growth in RNG consumption and in the number of domestic customers occurred between 1950 and 1972, when the cost of NG to residential customers was less expensive in many places than fuel oil and electrical energy.[6]

The number of RNG customers and RNG use per customer increased every year. The number of new RNG customers increased because NG was substituted for the inferior sources of energy for the home, such as coal and wood. The growth in the number of customers was supported by income growth, the cost of NG relative to the cost of competitive fuels, and the distinctive features of NG as a means of providing particular energy services for the household, such as cooking. Use per customer increased as customers continued to use NG more extensively within the home and more intensively in particular appliances as price declined. Consumption of NG by residential customers reached a peak in 1972, a year before the Arab oil embargo of 1973/1974.

Table 6.2

Natural Gas Use by Major Consuming Sector

	Major Consuming Sector				Residential Sector	
Year	Res	Comm	Indus	Elec	Cust	Rev
1949	0.99	0.35	3.08	0.55	14,690	666
1950	1.20	0.39	3.43	0.63	16,906	826
1960	3.10	1.02	5.77	1.72	31,148	3,902
1970	4.84	2.40	9.25	3.93	38,604	5,272
1972	5.13	2.61	9.62	3.98	39,871	6,224
%Change	418%	646%	212%	624%	171%	834%

Source: *Minerals Yearbook*, various years.
Note: Consumption is in trillion cubic feet (rounded). Customers (Cust) are in thousands of customers (rounded). Revenues (Rev) are in millions of dollars (rounded). Res is the residential sector primarily households. Comm is the commercial sector primarily businesses such as wholesale and retail stores. Indus is the industrial sector primarily manufacturing, mining, and manufacturing industries. Elec is electric utilities.

The growth in the consumption of NG between 1959 and 1972 was still greater for the industrial sector than for the residential sector. Although sales to the residential sector were gaining on the industrial sector between 1950 and 1960, sales to the industrial sector pulled ahead dramatically in the 1960's. This growth in industrial use was due to the especially low price at which NG was still available to industrial customers in the southern part of the United States, especially in the major producing states of Texas and Louisiana (see Tables 6.3 and 6.4).

REGIONAL PRICES AND THE GROWTH OF MARKETS

The average price of NG to industrial customers in Texas, Louisiana, and the other major NG producing states during the 1950's was low and not much greater than the well-head price. The situation was different in the North, especially in such states as West Virginia, Ohio, and Pennsylvania, where the industrial customer paid much more for NG. Yet, the domestic customer in the North still

Table 6.3

Basic Gas Industry Statistics in 1950

	Residential		Industrial		Wellhead		
	Use	Price	Use	Price	Price	Reserves	Production
Ohio	175	.60	112	.40	.19	659	43
California	179	.66	432	.19	.12	9,760	558
Pennsylvania	111	.67	130	.36	.25	627	91
Texas	81	.63	1705	.07	.05	102,404	3,126
Michigan	76	.85	41	.49	.13	195	11
Illinois	61	.98	157	.23	.10	230	13
Missouri	49	.70	79	.20	.14	a	h
New York	45	.89	10	.59	.25	65	3
Kansas	43	.48	178	.12	.07	13,791	364
Oklahoma	40	.45	229	.08	.05	11,634	482
West Virginia	35	.41	88	.25	.17	1,651	190
Minnesota	26	.72	37	.22	d	b	b
Kentucky	26	.56	28	.24	.20	1,331	73
Louisiana	24	.60	440	.08	.05	28,533	832
Colorado	22	.57	55	.15	.04	1,116	11
Nebraska	20	.66	34	.19	.12	44	h
Iowa	19	.69	40	.20	d	b	b
Indiana	17	1.11	37	.34	.07	31	1
Georgia	16	.67	45	.18	d	b	b
Utah[e]	15	.54	21	.21	.06	85	4
Arkansas	15	.51	113	.09	.04	908	48
Montana	13	.47	18	.14	.05	797	39
Tennessee	12	.73	51	.18	.10	b	b
D.C.[f]	12	1.30	3	.60	d	a	b
Alabama	12	.81	77	.18	.05	a	a
Mississippi	10	.70	61	.12	.06	2,519	114
Wisconsin	10	1.56	4	.71	d	2,195	b
Maryland	10	1.40	3	.74	.20	a	b
New Mexico	9	.66	134	.06	.03	6,991	213
Wyoming	6	.51	27	.09	.06	b	62
Arizona	5	.85	42	.20	d	b	b
Delaware[g]	3	.73	1	.37	d	b	b
Florida	1	.85	11	.15	.05	a	h
Total US	1,198	.69	4440	.13	. 07	185,593	6,282

Source: *Minerals Yearbook*, 1950, various pages.

Note: Use is consumption in billion cubic feet. Prices are in dollars per thousand cubic feet. Reserves are in billion cubic feet. Production is marketed production in billion cubic feet.

[a] Included in category of other states reserves which amounted to 27,472 million cubic feet. Some North Dakota reserves are also listed in this category.

[b] Indicates an amount less than 500 million cubic feet.

[d] Not applicable.

[e] Utah, South Dakota and North Dakota. North Dakota had 600 million cubic feet of marketed production with a well-head price of 5.1 cents per thousand cubic feet.

[f] District of Columbia and Virginia. Virginia had less than 50 million cubic feet of production with a well-head price of 8.7 cents per thousand cubic feet.

[g] New Jersey and Delaware. New Jersey had less than 50 million cubic feet of marketed production.

[h] less than 50 million cubic feet.

Table 6.4

Basic Gas Industry Statistics in 1972

	Residential		Industrial		Well-head		
	Use	Price	Use	Price	Price	Reserves	Production
California	637	1.08	624	.47	.37	5,329	487
Illinois	488	1.13	399	.64	.28	545	1
Ohio	478	1.05	429	.63	.39	1,147	90
New York	363	1.62	103	.86	.33	139	4
Michigan	355	1.10	272	.63	.31	1,297	34
Pennsylvania	306	1.36	365	.72	.30	1,407	74
Texas	241	1.01	1839	.26	.16	95,042	8,658
Indiana	169	1.14	291	.56	.15	87	b
Missouri	160	1.10	99	.44	.22	e	a
New Jersey	150	2.02	83	.81	e	e	b
Minnesota	107	1.23	105	.48	e	e	b
Wisconsin	105	1.34	138	.64	e	e	b
Kansas	101	.74	178	.33	.14	11,939	889
Iowa	97	1.11	105	.46	e	e	b
Colorado	89	.78	83	.33	.16	1,655	117
Maryland[d]	89	1.65	61	.75	.21	e	b
Kentucky	86	.93	78	.55	.25	938	64
Georgia	85	1.26	155	.53	e	e	b
Louisiana	83	.91	1017	.33	.20	74,971	7,973
Oklahoma	78	.90	125	.29	.16	14,492	1,807
Nebraska	60	.99	56	.49	.18	50	4
West Virginia	60	.96	94	.56	.30	2,346	215
Virginia	55	1.60	53	.60	.32	36	3
Tennessee	54	1.02	135	.50	.32	e	a
Alabama	53	1.27	165	.42	.35	246	4
Utah	49	.90	61	.36	.17	1,022	40
Arkansas	47	.83	146	.35	.17	2,456	167
Mississippi	39	1.08	147	.34	.27	1,104	104
New Mexico	35	.97	77	.34	.19	12,336	1,216
Arizona	34	1.24	63	.48	.18	e	b
Montana	24	.97	33	.38	.12	1,064	34
Wyoming	22	.74	55	.30	.16	4,089	375
South Dakota	13	1.13	6	.38	e	e	b
Florida	13	2.66	87	.50	.32	181	16
North Dakota	10	1.13	3	.49	.17	442	33
Delaware	8	1.71	10	.76	e	e	b
Total U.S.	5126	1.26	8169	.45	.19	266,085	22,532

Source: *Minerals Yearbook*, 1972.

Note: Use is in billion cubic feet. Price is in dollars per thousand cubic feet. Reserves are in billion cubic feet. Production is marketed production in billion cubic feet.

[a] Indicates less than 50 million cubic feet.

[b] Indicates an amount less than 500 million cubic feet

[d] Includes the District of Columbia.

[e] Not applicable.

paid about the same amount for NG as the domestic customer paid in Texas and Louisiana.

West Virginia residential customers were still enjoying the lowest price for NG in the nation in 1950. However, in Texas, a state with much greater production and reserves than West Virginia, the domestic customer in 1950 paid about $0.23 more per Mcf for NG. Domestic customers paid almost as much for NG in Texas as they paid in Georgia, a state which did not produce any NG and in which the nearest source of supply was several hundred miles away.[7]

Industrial customers, because of their size and because of their ability to locate in NG-producing areas, were able to obtain NG at a low price in 1950 and in later years. The relatively low price that industrial customers paid for NG encouraged not only NG use, but also industrial development and employment growth in Texas and in other south-central states. Thus, manufacturing employment increased between 1950 and 1960 from 363,500 workers to 489,500 workers in Texas, and declined in Pennsylvania and West Virginia.[8] Since manufacturing wages were less in Texas than in the northern states, part of the increase in employment was, of course, due to wage differences, but a part of the growth was due to the relatively low cost of NG.

REGULATORY INITIATIVES PRIOR TO THE NGPA

As the states of Texas and Louisiana were successfully pursuing an industrial policy which included low NG prices to industrial customers, the Supreme Court settled the Phillips Petroleum Company case in 1954, which effectively extended the control of the Federal Power Commission over NG prices to the well-head. This ruling was viewed favorably by many who thought that the wartime involvement of the United States Government in the energy industry should be continued, but unfavorably by the NG industry.[9]

Not too surprisingly, the oil and gas industry worked hard to reverse the ruling from the Phillips Petroleum Company case by legislative means. In fact, legislation to reverse the ruling was almost passed in 1956. However, President Eisenhower was forced to veto the deregulation bill because of reports that money was paid to Senator Francis Case of South Dakota by Superior Oil Company to influence his vote on the legislation. Failure to pass a deregulation bill further established the regulatory authority of the federal government in the NG industry, which had grown during the 1930's and the Second World War. Federal government involvement in the industry generally tended to maintain a steady cost of NG to residential customers.

During the Second World War the NG industry and the national government had worked closely both in planning and in developing the NG industry and in allocating NG to customers. After the war, studies appeared describing how this planning and development process should be continued in order to best serve the public interest.

Some consumer advocates even wanted to use a regulatory authority to ensure that adequate supplies of NG were available to consumers at least cost over the life of their NG using equipment. Senator Paul Douglas was a strong supporter of the need to control the price of NG from the site of the well-head to final distribution to consumers. Douglas claimed that not only was there too much control of a large portion of total reserves and production by a few companies, and not only were pipeline companies able to pass artificially high prices on to the consumer, but "escalation clauses, renegotiation clauses, and most favored nation clauses make these artificially high prices the new dominant or average prices. Thus, escalation clauses create a constant 'spiraling' of prices in which the previous high price becomes the new dominant price. To put it another way, the marginal or most recent and highest price becomes the average price."[10]

Douglas proposed regulating the activity of the 197 largest NG producing firms in 1954 that sold 90 percent of the NG purchased by interstate pipeline systems in that year, but not regulating the 5,360 producers who sold the rest. Douglas emphasized that such regulation was not only legal but it was also consistent with bureaucratic procedure.[11,12]

Despite the significance of NG service for residential customers, the major energy policy issues of the 1950's and 1960's were the protection of the domestic oil industry by restrictions on the importation of oil, the further development of the electric utility industry and, in particular, the nuclear electric industry and the effect of using different types of energy on environmental and workplace safety and health. NG use in the American economy continued to be addressed in policy debates, primarily because it created fewer environmental and occupational safety and health problems than other sources of energy, but not because it might supply households with the most efficient means of satisfying their heating needs.

CUSTOMER GROWTH

Between 1950 and 1960, the number of RNG customers almost doubled (see Table 6.5) and RNG consumption nearly tripled. The increase in consumption was due mainly to the large increases in the proportion of customers who used NG for space heating during the period.

The top six growth states in number of customers (see Table 6.5) only accounted for 33 percent of customer growth between 1950 and 1960 because the number of NG customers increased significantly in every part of the country. The growth in New York, however, was singularly significant.

The growth in New York was largely due to the conversion of MG customers to NG customers. Brooklyn Union Gas Company of Brooklyn, New York, alone

Table 6.5

Major Residential Growth Markets between 1950 and 1960

	Customers			Use per customer			Extent
	1950	1960	Change	1950	1960	Change	1960
California	2,682	4,252	58%	67	86	28%	85%
Illinois	1,607	2,177	36%	38	107	180%	71%
New York	704	3,786	337%	64	59	-7%	72%
Ohio	1,532	2,061	35%	114	176	54%	72%
Pennsylvania	1,633	2,045	25%	68	113	67%	61%
Texas	1,282	1,909	49%	64	90	42%	69%
US	16,906	31,148	84%	71	100	40%	59%

Source: *Minerals Yearbook*, various years.
Note: Customers are the number of residential customers in thousands. Sales per customer are in thousands of cubic feet. Extent is the number of residential customers relative to the number of occupied housing units expressed as a percentage. Change is the percentage change between 1950 and 1960.

converted 925,000 customers to NG in 1952. The conversion in Brooklyn from MG to NG, which was reminiscent of the conversion of 930,000 customers in Chicago twenty-one years earlier, required many changes to the distribution system,[13] and it also required the readjustment of every burner on every stove that previously had used MG. Service men were required to go into every house and make the necessary adjustments. The conversion saved the pipelines enormous amounts of money since the effective capacity of the distribution system was doubled because of the much higher Btu content of NG per cubic foot of gas sold. Twice as much useful heat could be dispatched along the same distribution channels when NG was substituted for MG. This increase in capacity encouraged distribution companies to add even more new customers.

The percentage reduction in residential gas customers' bills from the conversion to NG was generally less in the early 1950's than in the 1930's. Conversions were also greeted with much less enthusiasm by the public in the 1950's than in the 1930's since "people considered natural gas different. It cooked differently; the water wouldn't boil as fast, or it would boil too fast. Pots burned, people said. We had many, many legal claims as we went along: 'The pots are burned, the man dirtied my house, the canary died -it must have been the natural gas'."[14]

Illinois was also an important growth market after the second world war. The market in Illinois increased greatly from 27 percent of the housing units served with NG in 1940 to 59 percent in 1960. Between 1940 and 1960 sales per customer in Illinois grew seven-fold largely because of a large increase in the proportion of NG customers that were space heating customers.

By 1959 more than 90 percent of the gas customers were space heating customers in the West South Central, East South Central, Pacific, and Mountain census divisions of the United States (see Table 6.6). The East North Central census division experienced the greatest growth between 1949 and 1959 and between 1959 and 1969. The number of space-heating customers increased more than four-fold in this region between 1949 and 1959. Except for New England and Mid-Atlantic census divisions, growth after 1969 was minor since nearly all natural gas customers in all census divisions were space-heating customers by that time.

Table 6.6

Percentage of Space Heating to Total Gas Customers by Census Division

	1949	1954	1959	1964	1969	1972
West South Central	97%	99%	99%	99%	99%	99%
Pacific	87%	93%	97%	97%	96%	95%
Mountain	82%	94%	94%	97%	98%	98%
East South Central	56%	76%	86%	92%	96%	96%
West North Central	37%	67%	80%	88%	92%	94%
South Atlantic	29%	51%	64%	73%	80%	83%
East North Central	20%	41%	60%	76%	84%	88%
Mid-Atlantic	10%	21%	34%	42%	49%	52%
New England	4%	8%	32%	34%	46%	53%

Source: American Gas Association, *Gas Facts* (New York: American Gas Association, 1949-1970). American Gas Association, *Gas Facts* (Arlington, Virginia: American Gas Association, 1972).

Note: West South (W.S.) Central or WSC includes Texas, Oklahoma, Arkansas, and Louisiana. Pacific or PAC includes Washington, Oregon, and California. Mountain or MTN includes Montana, Idaho, Wyoming, Nevada, Utah, Colorado, Arizona, and New Mexico. East South (E.S.) Central or ESC includes Kentucky, Tennessee, Mississippi, and Alabama. West North (W.N.) Central or WNC includes North Dakota, Minnesota, South Dakota, Nebraska, Iowa, Kansas, and Missouri. South (S.) Atlantic or SA includes Florida, Georgia, South Carolina, North Carolina, Virginia, West Virginia, Delaware, Maryland, and the District of Columbia. East North (E.N.) Central or ENC includes Wisconsin, Michigan, Illinois, Indiana, and Ohio. Mid-Atlantic or MA includes Pennsylvania, New Jersey, and New York. New England or NENG includes Maine, New Hampshire, Vermont, Massachusetts, Connecticut, and Rhode Island.

USE PER CUSTOMER AND LOAD FACTORS

The significance of space-heating demand to overall levels of domestic NG demand can be observed by comparing the estimated average use of NG in space-heating units and in the other major appliances (see Table 6.7). In all parts of the United States, estimated use of NG for central space heating was still several times greater than its use for water heating and cooking. Use per space-heating unit also varied greatly by region of the country.

Table 6.7

Average Annual Natural Gas Consumption by Residential Appliance in 1966 (in 1000's of cubic feet)

Census Division	Space Heat	Water Heat	Gas Light	House Range	Clothes Dryer Gas	Clothes Dryer Electric	Gas Grill
NENG	133	23.8	19.0	9.8	8.9	4.8	2.8
ENC	131	28.0	18.4	10.2	8.5	4.5	2.2
MA	124	27.4	17.2	9.9	8.8	5.1	2.3
WNC	118	26.6	17.0	9.8	8.4	5.2	3.2
MTN	113	31.0	17.5	10.8	10.4	6.3	2.5
ESC	95	28.7	17.0	11.6	7.2	5.8	2.2
SA	88	23.4	18.2	8.9	8.7	4.8	3.0
PAC	73	27.0	18.2	11.5	9.3	4.5	3.3
WSC	71	23.0	18.1	11.9	8.2	5.4	4.1

Source: unpublished data, American Gas Association, Arlington, Va..

Note: The AGA statistics were converted to Btus using the conversion factor of 10.28 therms per thousand cubic feet. All amounts reported are for 1966 except Space Heat, which is from a similar 1975 survey. The 1975 data represent average consumption excluding extremes from 152 companies with a total of 28,079,339 residential customers or 69% of the 40,950,300 residential customers in 1975. The 1966 data represent approximately the same set of companies that reported in 1975. Gas or Electric Clothes Dryer indicates either a gas pilot or an electric pilot gas dryer. See Table 6.6 for definitions of Census Divisions.

Because of the relatively mild climate of the Pacific (in which California represented about 95 percent of total consumption in this division during the 1960's and the 1970's), space-heating units were estimated to use about 73 Mcf per year. Water heaters were estimated to use 23 Mcf. A customer in California who used NG solely for space heating and for water heating could be expected to use about 96 Mcf per year.

Although the amount of NG used to obtain heat for rooms in the winter time was clearly much greater than the amount of NG used for other purposes, the use of NG extensively for such purposes as water heating and gas lighting, however, had distinct advantages. The extensive use of NG within the household increased the load factor and, consequently, reduced the average cost of serving each RNG customer. In fact, if a large enough proportion of NG customers used NG for air conditioning and refrigeration, the load factor associated with serving residential customers exclusively would have reached a level between 60 percent and 70 percent in many places. Thus, the fact that the NG industry lost to the electric utility industry in the competition for the air-conditioning and refrigeration markets can be viewed as a lost opportunity for reducing the cost of NG to domestic customers.

In the late 1940's it was reported by the FPC[15,16] that when the load factor reached 60 percent, the cost of serving NG for many major pipelines was not much greater than when the load factor was 100 percent. This finding was motivated by a very poor presentation, based on hypothetical data, by the gas industry. The presentation attempted to justify the practice of selling NG to industrial customers at a discount. The findings in the presentation were found to overstate the effect that additional industrial customers had on increasing the load factor and on reducing the cost of delivering NG to all customers.

FPC economists in their presentation to counterbalance the industry presentation used actual cost and operating characteristics data from a variety of pipeline companies such as Colorado Interstate, Southern Natural Gas, Mississippi Fuel Corporation, and Panhandle Eastern. The FPC examined the differences in the load factor on the average delivered cost of natural gas. The FPC presentation showed that under a reasonable set of operating conditions for these pipeline companies, the industry greatly overstated the reduction in the cost of delivered gas from increases in the load factor and, in particular, the industry overstated the cost when the load factor was greater than 60 percent.

The industry had only considered the relationship between the load factor and the average cost for new pipelines which had high fixed costs. The industry had also not considered the influence of the average unit cost or variable cost of the gas at the well-head on the average cost of the gas delivered to market but had only considered the value of the gas that was required to run the compressors and other equipment in transporting the gas to market. The failure to consider such factors was shown to magnify the effect of increases in the load factor on the average cost of the gas delivered to markets. The industry also made other assumptions all of which tended to overstate the reduction in the average cost from increasing the load factor. In fact, the industry admitted in cross examination that it did not consider any factor, no matter how significant, if it tended to reduce the magnitude of the relationship between the load factor and the average cost of gas delivered to markets.

Why did the industry, as represented by the pipeline companies, overstate the size of the reduction in the cost of delivering NG from increasing the size of the load factor? One explanation is that the industry may have considered the own-price elasticity of the industrial customer to be much less than negative one in magnitude and several times as large in absolute magnitude as the own-price elasticity of the residential customer. Thus, the industry could maximize profits by attempting to create an environment where they would be able to offer NG to industrial customers at a very low price without fear of much public or regulatory disapproval. Regulatory bodies and other representatives of the public were apt to question rates and in some cases not approve proposed rates at rate hearings if the difference in the rates to be charged residential and industrial customers did not appear to be explained by differences in the cost of serving these two customer classes.

The industry case, however, was so poorly presented that it cast doubt on the industry's capability to manage its own economic affairs. Perhaps pipeline companies were selling NG to industrial customers at a price that did not cover the cost of service. Perhaps the industry presentation was clouded by conflicts of interests since many officials of pipeline companies also served on the board of directors of companies that they sold gas to.

Residential customers clearly had a need for an institution that represented their interest when legislation affecting the NG industry was being considered by the Congress. Yet no institution arose to serve this interest between 1950 and 1970 that was nearly as visible as the Cities Alliance had been during the 1930's. Instead, consumer magazines increased greatly in popularity and began to alert consumers to the advantages and disadvantages of obtaining energy from different sources for household services. These magazines, however, tended to emphasize the significance of this choice to the individual buyer and not to society as a whole.

CONSUMERISM AND NATURAL GAS APPLIANCES

The relatively low cost of NG as a source of energy within the household was being advertised by the gas industry in the 1950's at the same time as the dangers inherent in the careless use of NG within households was being discussed in consumer magazines. Although NG inhalation could not cause carbon monoxide poisoning, "the improper or incomplete combustion of this gas, as of any other fuel, can result in the production amounts of carbon monoxide. This was the case in the ten gas refrigerator deaths in New York City a year ago. Users of gas refrigeration have not generally been aware that the appliances require periodic adjustment and servicing."[17]

In addition to safety problems, the performance characteristics of gas refrigerators were also evaluated and were considered substandard. Servel was the major brand of gas refrigerator in the 1940's and the 1950's and the Servel refrigerator was incapable of maintaining a low enough temperature. For example, it was impossible to maintain the temperature of the Servel refrigerator below 59°F when the exterior temperature was 110°F. At 110° F, 46°F was considered to be the highest temperature for the refrigerator.[18] Even though gas refrigerators were generally much quieter than electric refrigerators, they were also more expensive and, along with gas air conditioners, they never became very popular. Whereas gas refrigerators were being panned, NG furnaces were generally receiving rave reviews.[19] NG furnaces were considered cleaner, more efficient, compact, and convenient, and even less expensive to maintain than other furnaces. Most safety problem associated with NG furnaces had been solved. Moreover, gas utilities in many places still serviced these appliances at little or no cost, which business practice tended to keep the furnace stock in a safe and efficient working order. Nonetheless, it was consistently pointed out that, whereas gas furnaces may have been relatively safe, they were not safer than electric space heating systems: "natural gas and LP gas present a special danger of explosion if they are not very carefully utilized. . . . A gas leak in the street can follow a water or drain pipe right into a cellar and has been known to do so on many occasions."[20]

Appliance manufacturers and the gas industry had also worked hard to secure a reputation for safety for the second largest application for NG in the household - water heating. Yet gas water heaters still required venting to the outside, and gas water heaters required a "'100 percent safety' shut-off control. Such a control functions to cut off the supply of gas to the main burner and the pilot in the event the pilot burner should go out, or in case of an interruption in the gas supply."[21] Gas water heaters were considered cleaner and more trouble free than oil burners. Gas also had the great advantage of heating water faster than electricity and a 30-gallon gas water heater was considered the equivalent of a 50-gallon electric heater. In areas where NG was available at a low rate, Consumer Reports considered NG water heaters to be a great buy.[22]

There were more similarities than differences among the ranges of the 1940's and the 1950's. In the 1950's, a 1940's style range could be purchased with the addition of a window in the oven door, a light within the oven, or a special griddle burner. Each addition increased the cost of the 1940's style range.[23] Electric ranges were being favorably compared with gas ranges, which greatly concerned the gas industry.[24] Although gas ranges were considered relatively safe, gas always posed a potential problem whereas electrical energy did not. Installation was also more difficult for a gas range than an electric range, and the gas valves and the air valves had to be carefully adjusted for proper performance. Electric ranges also gave off less heat than gas ranges, which kept a kitchen cooler during the summer. Electric ranges were also easier to clean. The operating knobs for the gas range tended to be on the front of the range where children could get at them. For electric ranges the knobs were likely to be above the range, out of reach for children. Yet Certified Performance ranges, which were much advertised by the gas industry for their stated performance standards, were still found not to be in compliance with advertised standards in many instances in 1950.[25]

As personal income rose and women remained in the work force after the Second World War, automatic tumbler dryers became very popular. Yet, in 1951 many gas dryers still required the lighting of a pilot light each time the dryer was turned on. Conversely, electric dryers, the competitive dryers, were fully automatic. Although the operating cost of a gas dryer was less, the initial cost of a gas dryer could be 15 percent to 20 percent greater than the initial cost of a similar electric dryer. The Sears Kenmore dryer had list prices of $199.95 and $239.95 for the electric and gas models, respectively.[26] Both dryers were generally considered safe. Yet, "With a gas dryer, venting to the outdoors is absolutely essential, to dispose of the poisonous products of combustion of the gas."[27] Electric dryers generally outsold gas dryers.

The increased purchase of NG furnaces and water heaters during the 1950's and 1960's contributed greatly to the increased growth in the overall NG market. The growth in the market also improved the quality of services within the home as many of these purchases were replacements for coal furnaces and heaters. However, the growth was to end in the 1970's as a consequence of several factors.

NOTES

1. U.S. Federal Power Commission, *Natural Gas Survey, Vol. 1, FPC Report* (Washington, D.C.: U.S. Government Printing Office, 1976), 29. By 1950 the technology was available to compress 1,000 cubic feet of NG at city main pressure to 26 cubic feet at 600 pounds pressure. This was similar to the compression that might take place in long-distance NG transmission.

2. J. R. Frey and H. C. Ide, *A History of the Petroleum Administration for War 1941-1945* (Washington, D.C.: U. S. Government Printing Office, 1946), 231-232.

3. R. W. Hooley, *Financing the Natural Gas Industry* (New York: Columbia University Press, 1961). This text describes the importance of the insurance industry in the 1940's and in the 1950's for the development of the NG industry. Life insurance companies were the major source of capital for every new, important pipeline system. Insurance firms also financed much production activity.

4. Several texts were cross-referenced in identifying the technological improvements. Clark, *Chronological History*. F. Mangan, *The Pipeliners* (El Paso, Texas: Guynes Press, 1977). American Gas Journal, *Diary*. Federal Power Commission, *Natural Gas Investigation (Docket No. G-580) Report of Commissioner Nelson Lee Smith and Harrington Wimberly*, (Washington, D.C.: U.S. Government Printing Office, 1948).

5. The average price of NG to residential customers in real dollars declined during the late 1940's, the first quarter of the 1950's and the 1960's when the commissioners of the FPC tended to be pro-consumer, and increased during the last three quarters of the 1950's when the commissioners tended to be pro-business. Sanders, *The Regulation of Natural Gas*, 94-124.

6. In 1960, NG was less expensive than fuel oil on a Btu equivalent basis in a majority of the states where both fuels were available. Electricity was more than eight times as expensive as NG on a Btu equivalent basis in 1960 in 23 states. Energy Information Administration, *State Energy Fuel Prices by Major Economic Sector from 1960 Through 1977*, DOE/EIA-0190 (Washington, D.C.: Energy Information Administration, July 1979).

7. The relatively low cost of NG to residential customers in Georgia in 1950 is explained by several factors. The pipeline system serving NG to Georgia, Southern Natural Gas, was a relatively efficient pipeline system. The fact that Eugene Talmadge was governor of Georgia several times between 1932 and 1946, when he died, is important. (His son was elected governor in 1948). Shortly after he was elected governor, Talmadge fired all six members of the public service commission because they refused to go along with his demands for a reduction in the price of NG. Prices were subsequently reduced. The fear of public denunciation and worse most probably encouraged the public service commission to keep the cost of NG to residential customers low during the 1930's and the 1940's. Tate, *Keeper of the Flame*, 133-135. V.O. Key, Jr., *Southern Politics in State and Nation* (New York: Alfred A. Knopf, 1949). C. M. Logue, *Eugene Talmadge, Rhetoric and Response* (New York: Greenwood Press, 1989).

8. Bureau of Labor Statistics, *Employment and Earnings, States and Areas, 1939-1975*, Bulletin 1370-12, (Washington, D. C.: U.S. Government Printing Office, 1977).

9. W. J. Barber, "The Eisenhower Energy Policy: Reluctant Intervention," in *Energy Policy in Perspective, Today's Problems, Yesterday's Solutions*, ed. C. D. Woodward (Washington, D.C.: Brookings Institution, 1981), 205-286.

10. P. H. Douglas, "The Case for the Consumer of Natural Gas," *The Georgetown Law Journal*, 44 (1956): 589.

11. Douglas, "The Case for the Consumer," 602-606.

12. Despite an overall strategy by the Eisenhower administration of disengaging the government from involvement in industrial decision making and the allocation of resources between consuming groups or regions, the FPC considered the policy during his administration of invoking eminent domain for the purpose of developing underground NG storage reservoirs. This policy would have worked to the advantage of northern residential consumers. Barber, "The Eisenhower Energy Policy," 224.

13. Problems encountered in switching a pipeline system from MG to NG varied. In some instances the NG would dry out the joints in cast iron mains and also dry out the leather diaphragms in meters which would cause leakage. Conniff and Conniff, *The Energy People*, 267-277. In other instances the switch to NG would result in the deposit of oil and moisture in mains. This could result in the formation of oxide scales in pipes which eventually fell off and could be carried to meters and cause problems there. Stotz, History, 296-298. If the NG was dry and the MG was wet, leakage of the NG could occur wherever a pipe was connected to another pipe or other equipment when the NG was substituted for the MG. Some companies rehydrated the NG with steam and oil fog before sending out the gas to households to make the NG similar to the MG. Tate, *Keeper of the Flame*, 97-113.

14. Conniff and Conniff, *The Energy People*, 273.

15. Federal Power Commission, *Natural Gas Investigation*, 255-267.

16. Stockton, Henshaw and Graves, *Economics of Natural Gas*, 177-186.

17. Consumer Union, *Consumer Reports* (New York: Consumers Union, January 1953), 40.

18. Consumer Union, October 1951, 439.

19. Consumer Union, August 1952, 483. Others point out that gas refrigerators failed in the market because of the poor management of the major gas refrigerator companies. Competition from General Motors, General Electric, and Westinghouse, which invested large amounts of money and talent in the development of electric refrigerators, also helped. The notoriously conservative gas utilities offered little support for the development of the gas refrigerator. Although gas refrigerators were originally considered technically more efficient and reliable, and were quieter than electrical refrigerators, they eventually lost out to electric refrigerators. R. C. Schwartz, *More Work for Mother* (New York: Basic Books, Inc., 1983), 128-150.

20. Consumers' Research, Inc., *Consumer Bulletin - Annual 1967*, (Washington, New Jersey: Consumers' Research, Inc., 1967), 165.

21. Consumers' Research, April 1952, 182.

22. Consumers' Research, January 1955, 13.

23. Consumers' Research, October 1952, 452-455.

24. Consumers' Research, September 1958, 463.

25. Consumers' Research, October 1952, 455.

26. Consumers' Research, April, 1951, 158 for price quotes. For a more extensive discussion of dryers in the 1950's see Consumers' Research, July 1957, 308-311.

27. Consumers' Research, Consumer Bulletin - Annual 1963/1964, (Washington: New Jersey, 1964), 154.

7

LACK OF GROWTH IN RESIDENTIAL
MARKETS AFTER 1973

Several quite different events significantly affected RNG markets after 1973. Restrictions on RNG space heating hookups, which began in the early 1970's after the large buildup in the number of space heating customers in the 1950's and 1960's, reduced the annual increase in the number of RNG customers. The Arab Oil Embargo of late 1973 and early 1974 had the effect of reducing the level of NG demand per customer.[1] During the cold winter of 1976/1977, temperatures were lower than they had been in more than fifty years, which greatly increased consumer expenditures on NG during that winter. Finally, legislation passed in 1978 reduced Federal involvement in NG price setting.

CHANGE IN THE NUMBER OF CUSTOMERS

The percentage of space-heating customers to total customers declined from 99 percent to 97 percent between 1969-1974 in the West South Central United States, the major natural gas producing region of the country. This decline was due to gas utilities not allowing new customers to use NG for space heating because of claimed insufficient reserves of NG. The percentage of new homes equipped with a gas furnace generally declined until 1978 (see Table 7.1), when the Natural Gas Policy Act (NGPA) was passed and RNG prices began to steadily rise. Not too surprisingly, an increasing proportion of the new homes were to be equipped with electric and non conventional sources of space heating energy during much of the 1970's.

The NG industry was unfortunately in the depths of its problems between 1976 and 1979, just when the housing industry was experiencing a great growth period (see Table 7.1). The NG industry began to relax its restriction on allowing new customers to use NG for space heating just when the housing industry began a major slump.

Table 7.1

Percentage of New Houses with a Natural Gas Furnace - 1971-1982

Census Region	1971	1972	1973	1974	1975	1976	1977	1978	1979	1980	1981	1982
Northeast	42	36	34	29	24	15	17	16	24	35	35	36
North Central	78	74	61	51	49	48	46	50	55	60	62	57
South	49	42	33	27	29	29	27	27	28	30	31	33
West	78	72	69	66	59	60	57	52	51	52	51	49
U.S.	60	54	47	41	40	39	38	37	39	41	41	40

Source: Bureau of the Census, U.S. Department of Commerce, *Characteristics of New Housing*, U.S. Construction Report-Series C25 (Washington, D.C.: U.S. Government Printing Office) various years.

Note: The Northeast (NE) includes the New England and Mid-Atlantic Census Divisions. The North Central (NC) includes the East North Central and West North Central Census Divisions. The South (S) includes the South Atlantic, East South Central and West South Central Census Divisions. The West (W) includes the Pacific and Mountain Census Divisions. See Table 6.6 for definitions of Census Divisions. The number of new houses fell from 1,174,000 in 1973 to 866,000 in 1975 and then rose to 1,369,000 in 1978 and fell after 1978 to 632,000 in 1982.

THE ARAB OIL EMBARGO OF 1973/1974

The Arab Oil Embargo of 1973/1974 shocked the American public. Energy problems joined environmental and urban problems as major social dilemmas for the American public. Society was troubled by how quickly the economic conditions of life, such as the availablility and price of gasoline, could be affected by external forces.

In the late spring of 1973, the Arab nations first threatened production cutbacks against such unfriendly nations as the United States and the Netherlands. On October 16, 1973, OPEC decided to raise the price of exported oil by 70 percent. By the end of November 1973, an embargo was firmly in place and stayed in place until March 18, 1974.

The passage of the NGPA in 1978 was motivated in part by the remembered consequences of the embargo of 1973/1974, during which period American citizens and their representatives responded publicly and in significant numbers for the first time since the 1930's to energy problems. Gasoline and other energy prices rose significantly. Many Americans also waited in line at service stations for gasoline.

The National Opinion Research Center (NORC) was conducting a telephone survey during 1973 and 1974 to find out how individuals were responding to the embargo in terms of their energy-using behavior. Interestingly, it was the first time the NORC was monitoring the public's reaction to a critical event as it developed. The NORC conducted a weekly telephone survey which was based on a national probability sample.

The NORC found that in four of the eight weeks between January 10, 1974 and February 28, 1974, the median daytime temperature within the home was reduced 3°F from 1973 winter levels of 68°F, a decline of approximately 4%. For the period from January 10, 1974 to February 28, 1974, 60 percent of the NG heated housing units reported turning down their thermostats.

Table 7.2

Percentage of Homeowners Initiating Particular Conservation Measures by Type of Heating Fuel after the Embargo of 1973-1974

Heating Fuel	Fixed Furnace	Changed Heating	Added Windows	Weatherized Windows	Closed off Rooms
Natural Gas	2.3%	2.9%	6.3%	9.4%	12.4%
Fuel Oil	1.3%	5.6%	7.8%	11.6%	15.5%
Electricity	1.1%	7.4%	6.4%	7.4%	27.7%
All types	1.8%	4.7%	6.5%	8.9%	15.0%

Source: J. R. Murray, W. J. Minor, R. F. Cotterman, and N. M. Bradburn, *The Impact of the 1973-1974 Oil Embargo on the American Household*, (Chicago, Illinois: National Opinion Research Center, 1974), 59.

Note: Changed Heating is the percentage of homeowners that changed the equipment they used for heating services. Added windows is the percentage of homeowners that added storm windows or doors.

The NORC also found that the American public was taking other actions in addition to adjusting thermostats in response to the embargo (see Table 7.2). Many citizens closed off rooms, put weatherstripping on windows, added storm windows/doors, and fixed furnaces. The embargo of 1973/1974 motivated all households to take action almost as much as it motivated those households using fuel oil, whose energy costs during the embargo rose most dramatically.

THE COLD SPELL BETWEEN SEPTEMBER 1976 AND JANUARY 1977

At the same time as the NG industry was unwilling to hook up new customers at prevailing prices, a shortage of another sort was developing in the gas industry.

The cold spell of 1976/1977, which began in September 1976 and extended through January 1977, was coupled with NG underground storage operators' not filling storage reservoirs up to planned levels. Then, operators withdrew NG out of storage reservoirs sooner than expected. This created a situation where the fear of shortage became a realization of shortages for some customers. The technology of the gas industry was unable to deliver NG to all customers, and some customers shipments of NG were curtailed. These curtailments were widely reported in the news media.

The cold spell of 1976/1977 affected every part of the country except the West Coast (see Table 7.3). This was a very unusual five-month period. Recorded data for heating degree days by the Census division indicates that heating degree days had never before exceeded 4000 heating degree days for a five month time period for New England, Mid-Atlantic, and East North Central census divisions. Since heating degree days are approximately normally distributed, the chance of observing values greater than an observed value can be readily calculated. The chance of observing a five month period as cold as or colder than September 1976 to January 1977 is between 1 in 500 and 1 in 1,000 for six of the eight census divisions.

Table 7.3

Heating Degree Days between September 1976 and January 1977

Division	NENG	MA	ENC	WNC	SA	ESC	WSC	MNT
Actual	4284	4078	4738	4732	2444	2386	2127	3306
Average	3580	3217	3578	3857	1758	2087	1446	3149
Deviation	20%	27%	32%	23%	39%	12%	47%	5%

Source: National Oceanic and Atmospheric Administration, *State, Regional, and National Monthly and Seasonal Heating Degree Days Weighted by Population* (1980 Census).

Note: See **Table 6.6** for definition of Census Divisions. The average is a fifty year average from 1931 through 1980. The deviation is the percentage deviation from the average.

The especially cold fall and winter season was particularly significant for domestic customers in the North since the extreme cold increased even further their normally high expenditures for NG service during the winter even further. Many middle-income and lower-income families needed to curtail expenditures on other items significantly during the winter holiday season because of the decline in income available for non-energy expenditures.

THE NGPA

Deregulation of NG prices was perceived by many as a means to increase NG supplies. Tax credits for conservation and other energy-saving investments were also considered to reduce demand.[2] Reduced demands for and increased supplies of NG would reduce the dependency of the United States on oil. It would also enable the United States to better satisfy both winter peak demands of the many households that used NG and demands of other households who wished to use NG but were unable to obtain it because sellers were unwilling to sell more NG to pipeline companies at prevailing prices.[3]

An increase in price for NG was not believed to greatly affect the conservation behavior of residential consumers. Thus, requirements for manufacturers to clearly label the efficiency of different appliances, and tax credits for different types of energy saving investments were being considered in additional legislation.

Hearings on the NGA addressed the monopoly powers of the gas industry, the authority of the national government in NG transactions, the possibility of consumer representatives negotiating directly with producers and pipeline companies for NG, and the institutional feasibility of transporting relatively inexpensive surplus NG from the South to the North. Congressmen at the hearings on the NGPA debated whether the government could craft a policy to increase exploration and production activity for the NG industry without undue cost to the rest of society. In summary, there were few similarities but many differences between the hearings of 1937/1938 and the hearings of 1977/78. One similarity was that NG was still considered a premium fuel for domestic customers, which should be preserved for them if at all possible. The use of NG in industrial boilers had replaced the production of gasoline and carbon black as the applications for NG that should be eliminated. The great economic and operational efficiency of NG in expeditiously supplying quality heat for household use was still accepted.

The national context of the debate in 1978 had broadened since the size of the market had expanded enormously. Almost every senator and congressional representative gave at least one speech at the hearings in 1977/1978. Government bureaucrats had a much smaller direct role and few were called on to testify at the hearings. Only one industry group - the Columbia Natural Gas System - and one consumer group - the Cities Alliance played major roles at the hearings of 1937/1938. In 1978 consumer groups included the Consumer Federation of America, Americans for Democratic Action, and the National Association of State Regulatory Commissioners. Industry groups in 1978 included the American Gas Association, the Natural Gas Supply Committee, Independent Gas Producers Committee, Independent Petroleum Association of America, and the Interstate Natural Gas Association. In 1977/1978, consumer representatives in NG-consuming states had diverse views as to the effect of deregulation on RNG price and on the supply of NG available to residential consumers, whereas in 1937/1938 there was much more of a shared viewpoint among supporters of consumers.

Senator Gaylord Nelson[4] of Wisconsin thought that deregulation would cost every man, woman, and child in Wisconsin an additional $21.58 per year for NG. A representative of Brooklyn Union Gas Company[5] of New York estimated that a typical space-heating customer would save $97 in 1978 if NG were deregulated, since NG would be substituted for synthetic NG (SNG) and LNG, both of which were more expensive than even higher priced NG under deregulation.[6] Thus, the substitution of low-cost NG for the high-cost MG was discussed at the hearings in 1938, whereas in 1978 the discussion shifted slightly to the substitution of the relatively low cost NG for the high-cost SNG and LNG.[7] However, in 1938 a large amount of MG was used, but only a small amount of SNG was used in 1978. The discussions of SNG suggest how particular problems and special interests instead of general problems and the general interest dominated the hearings much more so than in the 1930's.

A consequence of the diversity of interests and viewpoints was that general quantitative answers to the overall effect on energy markets and the economy, which were available from models for forecasting energy demands and supplies, were used to focus attention by supplying ready answers to complicated processes. It was, however, impossible to examine in sufficient detail within the Congress how the economic behavior of agents in the NG market were represented within the models. The models were used as a means of almost magically supplying quantitative answers, that is as "black boxes" and not as means for enlightment.

Models were used because congressional representatives had grown to expect such quantitative answers to problems of policy since the 1960's, when the government and industry hired more economists and operations research analysts than ever before, many of whom were trained to provide such answers. In fact, the administrators or the staffs of the various government energy analysis offices were frequently operations research analysts who were able to use results from probability, statistics, economics, and computer science to devise formal procedures or models for addressing real world policy problems.

The models were being used to evaluate the effect of a wide variety of policy options. However, the evaluations obtained from the models were viewed with skepticism by some. "Each side of the deregulation debate has aligned a number of arguments often numerical - which support their contentions. . . . These arguments are arranged in pairs, in different orders, with different emphasis, stacked on top of one another like a set of dominoes. The only problem is that each side plays with its own set of dominoes."[8] Yet the models represented the interaction of a wide number of factors and sectors in the American economy, and always reported aggregate bottom-line results for the energy economy, such as the supply and demand for NG. For these reasons they appeared more compelling to many than the historical series of Senator Proxmire.[9]

The forecasts from the models also agreed in certain instances, and this agreement suggested to many that all the models were to some extent reasonable. Therefore, the advocate or administrator was free to pick from among these equally believable models and to present the results which best supported his or her position. Moreover, the administrator who had control over the model soon learned that it was possible to "tune" the model until desired results were obtained. In this way the specifics of the model neither had to be changed nor examined to any great extent. "Analysts" who had the endurance to fiddle with the model until desired results were obtained were well paid.

Almost all models yielded forecasts of dramatic increases in the demand for NG by residential customers. Despite an expected increase in the real price of NG, NG demand was forecasted to increase.[10] This result was consistent with the common belief that there was a great "pent-up" or "latent" demand for NG that had not been satisfied because of government regulation. Thus, the logical response for many congressional representatives and private associations was to support any legislation which would change the regulatory apparatus to satisfy this demand.

CHANGES IN NATURAL GAS AND OTHER ENERGY MARKETS BETWEEN 1974 AND 1983

After 1973 the energy intensity of American society, as measured by end-use energy per capita, began to decline. From a level of 285 MMBtu per capita in 1973, it declined by more than 20 percent to 226 MMBtu per capita in 1983.[11] Between 1974 and 1983 the use of NG declined more than any other major source of energy for American society (see Table 7.4). Steady growth in electrification

Table 7.4

Consumption by Major Sources of Energy (in quadrillion Btu)

Year	Coal	Gas	Petroleum	Other	Nuclear	Total
1974	12.66	21.73	33.45	3.31	1.27	72.54
1977	13.92	19.93	37.12	2.51	1.70	76.29
1980	15.42	20.39	34.20	3.20	2.74	75.96
1983	15.89	17.36	30.05	3.87	3.20	70.50

Source: Energy Information Administration, *Annual Energy Review 1987*, DOE/EIA-0384 (Washington, D.C.: Energy Information, 1988), various pages.
Note: Gas is natural gas. Other is hydropower, geothermal and other sources of energy. Nuclear is nuclear electric power.

of American households, commercial establishments, and factories kept the use of hydropower, geothermal, nuclear electric, and coal growing during the time period, while the increase in the price of NG and petroleum kept the consumption of these fuels down. New regulations also encouraged the substitution of coal for NG.

The use of NG in all major consuming sectors was lower in 1983 than in 1974 (see Table 7.5). In the residential sector, consumption declined despite an increase in the number of customers and in the proportion of customers that were space-heating customers (saturation) between 1974 and 1983. The increase in saturation nationally occurred because the proportion of space-heating customers to total gas customers continued to increase in the East North Central and in the Middle Atlantic and New England Census divisions, where the price of fuel oil was greater than the price of NG.

Table 7.5

Natural Gas Use by Major Consuming Sector in 1974-1983

Year	Major Consuming Sector				Residential	
	Res	Comm	Indus	Elec	Customers	Revenues
1974	4.79	2.56	8.31	3.43	41,509	6,849
1977	4.82	2.50	6.82	3.19	41,366	11,324
1980	4.75	2.61	7.17	3.68	44,114	17,497
1983	4.38	2.43	5.64	2.91	45,153	26,564

Source: Energy Information Administration, *Natural Gas Annual 1986*, DOE/EIA-0131 (Washington, D.C.: Energy Information Administration) various pages.
Note: Consumption is in trillion cubic feet (rounded). Customers are in thousands of customers (rounded). Revenues are in millions of dollars (rounded). Res is the residential sector primarily households. Comm is the commercial sector primarily small businesses. Indus is the industrial sector primarily mining and manufacturing establishments. Elec is electric utilities.

The negative effect of the large increase in the real price of NG on use per customer was great enough to dominate the positive influence on use per customer of the increase in saturation (see Table 7.6). Nonetheless, the amount of revenues received by the gas utility industry from each domestic customer increased. Revenues increased four-fold because of the especially large increase in price and because the own-price elasticity was inelastic.

Between 1978 and 1983 when the price increase was greatest, revenues received from residential customers almost doubled. In real dollars, the increase in revenues was 39 percent. If use per customer had not declined between 1978 and 1983, revenues received from residential customers would have been $32 billion instead of the $26 billion received by the gas industry. Thus, price and deregulation mattered in three important ways: it significantly increased the amount of national income allocated to the NG industry; it prompted consumers to once more demonstrate their ability to conserve on the use of this resource; and it reduced the standard of living for many residential customers.

Table 7.6

Sales, Saturation, and Real Price of Natural Gas in the United States

	Sales	Saturation	Price	Revenues	Deflator
1974	115	83.5%	$1.43	$164.5	147.7
1977	117	85.2%	$2.04	$238.7	170.5
1980	108	87.8%	$2.20	$237.6	246.8
1983	97	88.7%	$3.00	$291.0	298.4

Source: Sales/Customer, Saturation, and Price, Energy Information Administration, *Natural Gas Annual 1986*, DOE/EIA-0131 (Washington, D.C.: Energy Information Administration, 1987); Deflator, Bureau of Labor Statistics, U.S. Department of Labor, CPIU-all items.

Note: Sales is consumption per customer in thousand cubic feet. Price is the price of natural gas per thousand cubic feet in constant dollars. Revenues is average revenues received per customer in constant dollars. Saturation is the percentage of residential gas customer that were space heating customers.

NOTES

1. J. R. Murray, M. J. Minor, R. F. Cotterman, and N. M. Bradburn, *The Impact of the 1973-1974 Oil Embargo on the American Household* (Chicago: National Opinion Research Center, 1974).

2. The effect of these investments on NG use was examined in a regression analysis of RNG demand per customer for six East North Central states in Herbert, "Demand for Natural Gas at the State Level," 79-87. The elasticity, although significant at the 5% significance level, was less than .05 in value.

3. P. W. Mac Avoy and R. S. Pindyck, *Price Controls and the Natural Gas Shortage* (Washington, D. C.: American Enterprise Institute for Public Policy Research, 1975).

4. U.S. Congress, *Congressional Record, House*, H8389, August 3, 1977.

5. U.S. Congress, *Congressional Record, Senate*, S16315, October 4, 1977.

6. U.S. Congress, *Congressional Record, House*, H8219, August 1, 1977.

7. In a SNG unit, a petroleum feedstock like naptha is made to react over a nickel catalyst with steam and hydrogen to yield a high-Btu gas that, like NG, consists mainly of methane. It is sometimes enriched with propane to enrich its Btu content to the level of NG.

8. U.S. Congress, *Congressional Record, Senate*, S16321, October 4, 1977.

9. U.S. Congress, *Congressional Record, Senate*, S16311-S16317, October 4, 1977.

10. W. H. Babcock, S. B. Siegel, and C. Swanson, *Energy Demands 1972-2000* (McLean, Virginia: PRC Energy Analysis Company, PRC-D-2017, May 1978). For a detailed commentary on the model and the forecasts used and produced by the government just prior to the passage of the NGPA see M. Holloway, ed., *Texas National Energy Project, Part II* (Austin, Texas: Texas Energy Natural Resources Advisory Council, 1979). The governments forecast for 1985 for the residential sector demand was 5.3 Tcf which was lower than many other forecasts. The actual level of demand in 1985 was 4.4 Tcf.

11. Energy Information Administration, *Annual Energy Review 1987*, (Washington, D.C.: Energy Information Administration, 1988), 43.

8

QUANTITATIVE ANALYSIS FOR
THE PERIOD AFTER 1960

Participants at the NGPA hearings generally assumed that RNG use per customer was not much affected by changes in the price of NG. Forecasts of the demand for NG were optimistic. Consequently, the decline in use per customer that occurred after large price increases in 1979 and later years surprised many analysts.[1] This same assumption about the effect of price on demand also influenced the decision to allow energy conservation investments to be used as tax credits by consumers. Specification of a neglible effect of price on RNG use per customer was consistent with the findings of the economist Piero Balestra in an influential applied econometrics book published in 1967[2] and in a paper coauthored with Mark Nerlove.[3] In these works it was found RNG demand per customer to be very price inelastic and essentially equal to zero in the short-run.

Information concerning price effects was increasingly used in the 1970's to influence government and private decision making that, in turn, affected the development of NG markets. At the NGPA hearings such information was used to evaluate the effect of increases in the price of NG on the economic welfare of gas consumers.

Since regression analysis was the most common method of estimating price effects, new results from such analysis for the time period as well as results from previously reported analyses are considered here. The historical data is examined in a way that might be prescribed as a general method for examining historical data prior to estimating a regression equation. Estimated price effects are also used to examine some decisions made by the gas industry during the time period that influenced the type of market development that occurred.

RNG consumption per customer rather than penetration of NG markets or the proportion of households in a state that use gas is examined here. This is done for several reasons. First, important data for analyzing market penetration are not available. For example, it is necessary to have information on what proportion of the cities throughout a state had access to a gas pipeline and how this proportion changed over time. Second, it is necessary to have information on whether the pipelines were of sufficient capacity, and whether the pipeline companies had contracts with suppliers for sufficient amounts of gas, to serve all customers that might demand NG. For example, Mississippi Fuel Pipeline Company was serving St. Louis NG in the 1930's. But it was not until the close of the 1940's that Missippi Fuel Pipeline Company entered into contracts for sufficient NG to serve all customers that demanded it. Third, it is necessary to have information on the exact type of restrictions that utilities had on adding new space-heating customers during the 1970's.

Previous studies of use per customer are greater in number, are similar, and generally yield more reliable estimates than studies that have examined fuel choice. Thus, results reported here and results reported in previous studies can be compared.

A case study is first presented in which some additional background information on the evolution of RNG and other energy markets after 1950 is reviewed. Then, the effect of the price of NG and other factors on RNG demand per customer during the 1960's and 1970's for a selected set of states is examined. The choice of states is selective in order to further simplify the model employed in the regression analysis. This examination includes a reinvestigation of some results reported by Balestra in 1967 and of estimated price effects from previous studies.

BACKGROUND FOR A CASE STUDY

In 1950 coal and wood were still widely used for space heating especially in states east of the Mississippi River - wood in the South and coal in the North. Fuel oil was widely used for space heating in New Jersey and in the upper northeastern and northwestern parts of the country and in upper Michigan and Wisconsin near the Great Lakes. By contrast, in states such as Oklahoma, Louisiana, New Mexico, and Texas, where residential customers lived near the major NG and oil wells, gas was used as the primary space-heating fuel in more than half the households. By 1960 more than half the households in the states bordering this region, such as Arkansas and Mississippi, used NG as their primary space-heating fuel. The average levels of use per customer in this new region began to approach levels to be found in western Pennsylvania, eastern Ohio, and West Virginia in the second decade of the twentieth century.

In 1960, 43 percent and 32 percent of the households in the United States used gas and fuel oil, respectively, as their primary space heating fuels. Some of this gas was still MG. By 1970 these percentages were 50 percent and 22 percent for NG and fuel oil, respectively. Between 1960 and 1970 a constantly increasing proportion of households were using NG and electrical energy in place of fuel oil for space heating. However, in all states the primary space heating fuel was still either NG or oil (see Table 8.1).

Table 8.1

Space Heating in Households in 1970 (as a percentage of total households)

	Gas	Oil	LPG	Electricity	Other
Alabama	55	1	15	4	25
Arizona	81	0	9	10	0
Arkansas	59	1	24	2	14
California	90	0	2	7	1
Colorado	90	0	2	7	1
Connecticut	36	53	2	6	3
Delaware	24	60	3	4	9
Washington, DC	58	28	1	7	6
Florida	16	34	22	16	12
Georgia	58	2	16	5	19
Idaho	43	30	5	4	18
Illinois	57	13	3	4	23
Indiana	70	19	2	3	6
Iowa	87	6	2	2	3
Kansas	91	1	3	4	1
Kentucky	72	3	4	4	17
Louisiana	83	0	10	3	4
Maine	6	91	2	0	1
Maryland	37	50	3	5	5
Massachusetts	33	59	2	3	3
Michigan	83	7	2	3	5
Minnesota	78	12	2	5	3
Mississippi	50	0	26	3	21
Missouri	83	6	4	2	5
Montana	73	0	5	0	22
Nebraska	88	5	1	4	2
Nevada	34	2	3	59	2
New Hampshire	32	68	0	0	0
New Jersey	36	54	2	3	5
New Mexico	90	0	5	5	0
New York	36	55	2	2	5
North Carolina	18	50	7	6	19
North Dakota	49	32	5	14	0
Ohio	89	2	1	3	5
Oklahoma	82	0	11	4	3
Oregon	31	47	3	15	4
Pennsylvania	52	36	1	2	9
Rhode Island	39	55	2	2	2
South Carolina	24	32	14	3	27
South Dakota	65	17	17	0	1
Tennessee	55	3	5	19	18
Texas	81	0	11	4	4
Utah	88	0	3	9	0
Vermont	16	84	0	0	0
Virginia	25	49	3	5	18
Washington	37	36	1	22	4
West Virginia	62	7	0	3	28
Wisconsin	70	20	1	4	6
Wyoming	98	0	0	0	0

Source: Bureau of the Census, *Census of Housing 1970*.

Note: Other/None includes coal or coke, wood, other fuel, none. Gas is utility gas. Oil is fuel oil, kerosine, and other petroleum fuels. LPG is bottled gas, tank gas, or LPG.

Probably nowhere else in the country was NG used as extensively within the household as in the south-central states, particularly New Mexico, Texas, Oklahoma, Arkansas, Louisiana, and Mississippi. Most of these states had only a few major cities in which a large proportion of NG customers lived and the majority of NG customers within each of these states experienced similar weather, paid a similar price for NG, used NG equally widely, and lived within a few hundred miles of each other. Important conditions influencing NG use per customer within each of these states were quite homogeneous. Texas was the exception. Houston, Austin, Dallas, and El Paso in the southeastern, central, north-central and western portions of this vast state, respectively, each had more than 200,000 customers nearby; each were quite different. There were differences in the weather, in the NG price, and in other factors that could affect NG use per customer. To properly analyze RNG use within Texas would require an analysis of NG use in each of these different parts of the state. Therefore, Texas is not treated in detail.

Although conditions under which residential customers used NG were similar within a state, there were differences among states. In areas within Lousiana and Oklahoma where NG was not generally available, LPG was widely used as a space-heating fuel, while in Arkansas and Mississippi wood was more likely to be used instead of LPG in such places in 1960. Such differences indicate the very regional character of energy use which still existed in 1960.

Coal was used very little as a space-heating fuel in Mississippi in 1960, whereas it was used widely in the neighboring state of Alabama. Alabama was a major coal mining and marketing area, and coal was usually more readily available and cheaper in Alabama than in Missippi. Moreover, since per-capita income was higher in Alabama than in Mississippi, many more households were likely to have converted to coal from wood when space-heating units became more generally available and affordable between 1920 and 1950. In Arkansas only 1.7 percent of the households used an oil-based fuel as the primary source of energy for space heating in 1960. Directly to the north, in Missouri, 17 percent of the households used an oil-based fuel for this purpose.

Missouri, however, had two major areas where most of the population lived - the St. Louis area and the Kansas City area. NG was readily available since the mid-1920's in the Kansas City area. But, only MG and mixed gas were available to many households as late as 1950 in St. Louis. At that time, Mississippi Fuel Pipeline, the NG supplier to the area, negotiated for receipt of large amounts of NG from producers.

St. Louis was also a major fuel oil market. During the first half of the twentieth century, a very active oil transport business had developed along the Mississippi River. Large amounts of oil and oil products were shipped by barge transport from the oil-producing areas in the south to St. Louis and other ports. There were also major refineries and oil pipelines in the vicinity of St. Louis.

Once NG became available in 1950 in St. Louis and in the surrounding counties, the number of households that used NG as their primary space-heating fuel began to increase and increased several fold between 1950 and 1960. Despite this growth, the percentage of space-heating customers to total gas customers in Missouri in 1955 was still only 60 percent, compared to nearly 100 percent in Arkansas.

The states of New Mexico, Arkansas, Oklahoma, Mississippi, and Louisiana are interesting for analytic purposes because a relatively constant 99 to 100 percent of RNG customers in these states were space-heating customers between 1961 and 1981. Had the proportion of NG space-heating customers to total customers changed dramatically between years or increased consistently, it would have been necessary to consider whether the average efficiency of the space-heating stock was also changing, since newer furnaces tend to be more efficient than older furnaces.

There was also very little fuel oil use in these states during the time period. This also simplifies the estimation because the effect of the price of fuel oil on NG consumption per customer need not be considered. This is particularly appealing because state values for the price of oil are usually measured very poorly.[4]

PARTICULAR FACTORS AFFECTING USE PER CUSTOMER

Several factors are examined for their influence on NG use per customer: heating degree days (HDDs), NG price, and electricity price.[5] HDDs are generally a very good variable for explaining NG use per customer when a large proportion of the gas customers use NG for space heating. HDDs are equal to zero if the temperature is greater than some base value, usually 65° F. If the temperature is below the base value, it is computed as the difference between the base value and the average of the high and low temperature for the day. Annual values are obtained by taking the sum of the daily values. HDDs are commonly used by many utilities in estimating RNG demand. In states where a large proportion of NG customers use NG for space heating, the strength of the relationship between HDDs and use per customers is so strong that the states with highest and lowest average levels of use per customer correspond to states with the highest and lowest average levels of HDDs (see Table 8.2).

The expectation is that the higher the price of electricity in a state relative to the price of NG, the greater the number of NG appliances purchased by NG customers, and the greater the amount of NG used by each RNG customer. Sellers of electric and NG appliances were competing for market shares during the 1960's and the 1970's in the south-central United States. It is, however, difficult to discern from mere observation of the data a general relationship between the state values for the average price of electricity and the average level of use per customer.

Table 8.2

Average Residential Consumption per Customer and Related Statistics

State	Use	Gas price	Degree days	Electricity price
New Mexico	122(14)	.78(.19)	4756(249)	6.51(1.04)
Arkansas	117(11)	.68(.09)	3353(237)	5.95(1.38)
Oklahoma	115(8)	.71(.10)	3590(248)	5.87(1.49)
Mississippi	93(6)	.86(.18)	2583(223)	4.74(0.79)
Louisiana	86(6)	.75(.19)	1853(212)	5.29(1.22)

Sources: Natural Gas Consumption and Gas Price, American Gas Association, *Gas Facts* (various years); Heating Degree Days, National Oceanic and Atmospheric Administration, *State, Regional, and National Monthly and Seasonal Heating Degree Days Weighted by Population*; Electricity Price, Edison Electric Institute, *Statistical Yearbook of the Electric Utility Industry* (New York: Edison Electric Institute, 1961-1977), *Statistical Yearbook* (Washington, D.C: Edison Electric Institute, 1978-1981).

Note: The data presented here are averages for the period from 1961 to 1981. Use is natural gas consumption per customer in thousand cubic feet. Both Gas price and Electricity price are in constant dollars per thousand cubic feet. Degree days are heating degree days similar to the statistics reported in Table 7.3.

Examining averages across states provides some indication of the nature of the relationship between RNG demand per customer and price. Examining the same data within a state over time (see Table 8.3) provides additional information.[6] A rise in NG prices after 1977 was associated with a decline in NG consumption per customer in all states. In 1973, when the cost of NG per MMBtu was $0.64 in New Mexico, NG consumption per customer was 133 MMBtu. By 1981, when the cost of NG in constant dollars was 88 percent more expensive, the level of demand was 35 percent lower. Similar relationships can be observed for other states. When prices declined between 1961 and the early 1970's, NG use rose in all states except New Mexico. Thus, the expected negative relationship between NG use and price of NG is directly observable from data.

Examination of price of electricity and NG consumption per customer, however, would still not suggest a relationship between the two series.

APPLIANCE USE

An indication of the degree of competition for NG and electric service within households can be obtained by examining changes in fuel choices for the different types of energy equipment used in households over time. The proportion of

Table 8.3

Residential Gas Consumption per Customer and Related Statistics

	New Mexico			Arkansas			Oklahoma			Louisiana		
Yr	GD	GP	EP	GD	GP	EP	GD	GP	EP	GD	GP	EP
61	136	.68	8.39	118	.70	9.13	111	.69	8.92	84	.66	7.67
62	130	.69	8.30	120	.70	8.77	118	.68	8.53	85	.68	7.36
63	127	.69	8.06	116	.69	8.17	113	.68	8.11	89	.67	7.15
64	143	.67	7.75	121	.70	7.69	113	.71	7.71	93	.67	6.85
65	134	.67	7.41	114	.70	7.24	116	.71	7.38	82	.67	6.52
66	139	.65	7.16	122	.68	6.75	111	.70	6.98	90	.66	6.17
67	118	.73	6.92	118	.66	6.42	112	.68	6.72	84	.65	5.91
68	129	.70	6.62	130	.62	5.90	123	.64	6.31	95	.62	5.51
69	125	.69	6.31	131	.59	5.43	125	.65	5.84	93	.60	5.12
70	130	.64	5.91	133	.59	4.95	127	.61	5.38	96	.57	4.78
71	126	.63	5.65	125	.59	4.79	123	.60	5.16	87	.59	4.60
72	125	.65	5.66	122	.58	4.76	116	.61	4.98	80	.64	4.52
73	133	.64	5.34	120	.58	4.39	130	.58	4.53	87	.63	4.33
74	121	.68	5.11	109	.63	4.64	110	.60	4.19	76	.65	4.33
75	127	.75	5.27	114	.62	4.71	121	.63	4.14	82	.73	4.01
76	116	.82	5.41	115	.65	5.18	109	.74	4.56	87	.80	4.07
77	108	.98	5.73	114	.71	5.45	114	.85	4.85	88	.94	4.12
78	101	1.08	6.24	114	.76	5.29	117	.85	4.85	89	1.01	4.05
79	109	1.08	6.47	110	.78	4.91	116	.85	4.57	85	1.05	4.07
80	99	1.16	6.50	96	.83	5.36	104	.84	4.64	77	1.10	4.75
81	87	1.20	6.49	87	.92	5.09	92	.92	4.88	73	1.22	5.24

Source: Gas Consumption per Customer (GD) and Gas Price (GP), American Gas Association, *Gas Facts*, various years; Electricity Price (EP), Edison Electric Institute, *Statistical Yearbook of the Electric Utility Industry*, various years; Deflator, CPIU-all Items.

Note: Electricity prices in kilowatt-hour were converted to cubic feet by using the conversion factors of 3,412 Btu per kilowatt-hour and 1,029 Btu per cubic foot.

households using NG and electricity for space heating, water heating, and cooking changed between 1960 and 1980 (see Table 8.4). The increased affordability of these types of energy services enabled many households to have automatic water heaters for the first time and to use gas and electricity instead of wood for space heating.

The proportion of occupied housing units using electricity for space heating, water heating, and cooking increased consistently from 1960 to 1980. On the other hand, more housing units in all states (except Mississippi) between 1960 and 1980 used NG for space heating, water heating, and cooking. Electricity, instead of

NG, was the primary source of energy for water heating and cooking in 1980 only

Table 8.4

Energy Choices for Major Energy Services in Households (as a percentage of total households)

State		Space heating			Water heating			Cooking			Dryer[c]	
		1960	1970	1980	1960	1970	1980	1960	1970	1980	1960	1970
NM	Gas	70	78	74	62	73	73	52	51	48	2	9
	Elec	b	2	8	10	10	12	22	33	38	8	27
	LPG	12	10	10	9	10	12	13	11	10		
	Wood	10	5	6	1	b	a	10	5	a		
	None	1	b	b	16	7	2	b	b	b	90	64
AK	Gas	55	65	58	46	60	55	49	52	42	2	9
	Elec	1	4	16	5	10	26	13	22	39	4	23
	LPG	16	21	16	11	17	17	23	23	18		
	Wood	24	8	10	b	b	a	12	2	a		
	None	b	b	b	37	13	2	b	b	b	94	67
LA	Gas	79	82	69	65	78	68	74	73	60	6	19
	Elec	b	6	22	2	8	24	8	16	33	6	22
	LPG	11	10	7	6	7	6	13	10	7		
	Wood	7	2	2	b	b	a	4	b	a		
	None	1	b	b	16	9	1	b	b	b	88	58
MS	Gas	49	52	46	37	37	39	40	37	30	4	2
	Elec	2	11	23	10	26	43	20	36	50	1	24
	LPG	17	27	21	9	14	14	18	23	19		
	Wood	26	9	4	b	b	a	19	4	a		
	None	b	b	b	44	20	4	b	b	b	95	74
OK	Gas	77	79	69	72	76	68	68	61	46	4	12
	Elec	b	5	16	2	9	21	12	26	44	6	27
	LPG	15	14	10	11	12	10	17	13	9		
	Wood	5	2	4	b	b	a	2	2	a		
	None	1	b	b	15	4	1	b	b	b	90	61

Source: Bureau of the Census, U.S. Department of Commerce, *Census of Housing 1960*, 1970, and 1980.

Note: LPG is liquefied petroleum gas and other bottled gas. Gas is utility gas. Elec is electricity.

ᵃ In 1980 wood was no longer listed as a separate category for water heating and cooking.

ᵇ Less than 1%.

ᶜ Only statistics for 1960 and 1970 were reported for clothes dryers.

in Mississippi. Bottled gas, such as propane and LPG, were still used widely for household energy services in several states. Bottled gas like NG was convenient to use for household applications, yet it was usually more expensive. Bottled gas was principally used in rural areas where NG was not available.

One significant change for households was the increased use of clothes dryers. In 1960 only a small percentage of households had clothes dryers (see Table 8.4). Clothes were still generally dried outside the household. By 1970, however, 39 percent of the housing units had a clothes dryer. Yet the largest proportion of these dryers were electric, not NG. The relatively small increase in the proportion of households having NG clothes dryers in most states (Louisiana is the exception) indicates that any increase in the average use of NG within the household from the increased use of NG clothes dryers was probably quite small.

In summary, a larger percentage of households that were choosing either an electric or a NG water heater, cooking unit, or furnace for the first time were choosing the electric equipment. Thus, aggregate electricity use was rising at a rate greater than aggregate NG use. However, it is shown in the next section that NG use per gas customer was probably not changing much during the time period because of any change in the percentage of NG customers that were choosing electric equipment over NG equipment.

CHANGES IN NATURAL GAS USE PER CUSTOMER FROM CHANGES IN APPLIANCE OWNERSHIP

The proportion of NG customers who were also space-heating customers was relatively constant and near one in value during the time period. Thus, it can be assumed that NG use per customer did not change during the time period because of changes in this proportion.[7] However, between 1960 and 1970, and 1970 and 1980, an increasing proportion of NG customers were using NG for water heating and a decreasing proportion were using NG for cooking (see Table 8.5). Yet, an analysis of overall demand between these decades reveals that because of changes in these proportions NG use per customer scarcely changed. The relatively small size of the change in demand from a change in appliance ownership of the average customer is revealed by comparing the estimated percentage change in demand from a change in appliance ownership (see last two columns of Table 8.5) with the actual percentage change in overall demand between decades. NG consumption per customer is observed to have declined by 28 percent, 24 percent, 20 percent, 18 percent, and 15 percent in Arkansas, New Mexico, Louisiana, Mississippi, and Oklahoma respectively, between 1970 and 1980.

The estimates reported in Table 8.5 on the change in demand from changes in appliance ownership of the average NG customer suggest an interesting result. For the states examined here any change in NG demand between decades by the average NG customer (in response to changes in the price of NG and other factors) is probably due more to modifications in energy-using behavior in the household rather than to changes in the relative number of space heating, water-heating, and cooking units owned.

Table 8.5

Change in Natural Gas Use from Change in Appliance Ownership

State	Water Heating			Cooking			Change in Use	
	1960	1970	1980	1960	1970	1980	(60-70)	(70-80)
AK	84%	92%	95%	89%	80%	73%	+0.7%	+0.1%
LA	82%	93%	99%	93%	89%	86%	+2.0%	+1.3%
MS	75%	78%	84%	82%	70%	66%	-0.3%	+1.3%
NM	90%	93%	98%	75%	65%	64%	-0.2%	+1.3%
OK	93%	96%	97%	88%	77%	66%	-0.6%	+0.9%

Sources: Water Heating and Cooking, Bureau of the Census, *Census of Housing 1960*, 1970, and 1980, various pages; Number of Residential Customers, American Gas Association, *Gas Facts*, 1960, 1970, and 1980, various pages.

Note: The percentages reported in the last two columns of the table are derived by combining energy demand per appliance statistics from the American Gas Association (AGA) similar to those reported in **Table 6.7** with percentage of customers using natural gas for water heating and cooking reported above for each census year. These percentages are calculated as the ratio of the number of housing units that use utility gas as their primary water heating or cooking fuel as reported in the Census of Housing and the number of residential customers as reported by the American Gas Association. It was assumed, based on an examination of AGA annual data, that the percentage of space heating to total gas customers was constant between 1960 and 1980 and that the average efficiency of appliances did not change much during these years.

REGRESSION ANALYSIS

It is always difficult to summarize the historical relationship between NG consumption per customer and average price of NG and other variables with available data by means of an estimated regression equation. A major question is whether NG consumption per customer could be systematically related to some variable for which measurements are not available. For example, a large increase in the availability of NG within a state during the time period would suggest that there were customers living in parts of the state who were not receiving NG at the beginning of the period. Suppose these new households were different in ways that tended to increase average RNG demand per household; for example, if a larger proportion of the new customers resided in older homes or detached homes, had larger families or lived in larger homes; the average NG demand per customer might then increase during the time period. If such a change were systematically related with one of the included variables, then the estimated regression coefficients for the included variables would be biased. Because the availability of NG within the chosen states did not change much during the time period, the size of the bias in estimated coefficients from such a change is probably very small. However, there are other possible problems to examine.

An examination of national statistics, magazine advertisements, and articles from the 1960's through the 1980's suggests that there was an increase both in recreational and entertainment activities outside the household and also an increase in the proportion of households in which the husband and wife both worked away from home during the day. Naturally enough, as individuals spent less time in the household, they used less energy there, other things being equal. In a period of rising prices such as the 1970's, the estimated effect of the price of NG on NG consumption per household will tend to be overstated when such lifestyle variables are not accounted for in a regression analysis. However, the data (see Table 8.3) indicates a falling or relatively constant average price during the 1960's and only a rising price during the 1970's and early 1980's. Thus, the correlation between such lifestyle variables and the price of NG is probably small, and any bias in the estimated NG price coefficient from not including such variables is also small.

After having considered the several estimation issues discussed above and after having investigated previous econometric analyses, it was decided to estimate the following model by ordinary least squares:

$$GD_t = a - bGP_t + cHR_t + dEP_t - gD1974 + e_t$$

where,

GD	= RNG demand per customer,
GP	= average RNG price in constant dollars,
HR	= an indicator of space heating requirements per customer (see Appendix III)
EP	= average residential electricity price in constant dollars,
D1974	= an indicator variable for the year 1974,
e_t	= error terms assumed to be random and independent of each other over (t) and of the explanatory variables and to have constant variance,
t	= 1961 to 1981.

Expected signs are indicated in the equation, and all variable values are expressed in logarithmic form, and thus, the estimated regression coefficients can be interpreted as elasticities.

An indicator variable for the year 1974 (D1974) is included in the equation since examination of information from surveys and previous regression analyses suggest that households reduced their NG use in 1974 apparently in response to such stimuli as increased attention to energy issues by the news media after the Arab Oil Embargo, which began in late 1973. It is estimated that there was a shift downward in NG demand per customer of between 4 percent and 8 percent during 1974 in all states except New Mexico (see Table 8.6).

Table 8.6

Estimated Regression Coefficients, Standard Errors of Coefficients, and Other Summary Statistics (standard errors in parenthesis)

	GP	HR	D1974	EP	R^2	DW
Arkansas	-.85(.050)	+.53(.094)	-.06(.027)	+.17(.028)	.94	1.51
New Mexico	-.48(.043)	+.73(.106)		+.14(.054)	.97	1.79
Oklahoma	-.46(.065)	+.58(.131)	-.08(.041)		.79	2.31
Mississippi	-.34(.031)	+.47(.069)	-.06(.026)		.90	1.51
Louisiana	-.27(.028)	+.51(.060)	-.04(.029)		.89	1.69

Note: After an ordinary least squares estimation was used to obtain a demand relationship for New Mexico, an adjustment in the estimation for first order serial correlation was made using the maximum likelihood procedure in the Statistical Analysis System Proc Autoreg. The results of this latter estimation are reported here. The estimated coefficients for the adjusted and unadjusted estimates are very close in value. The results of both estimations as well as other estimated results observed in completing this analysis are reported in Appendix I of this chapter. Durbin-Watson (DW) is the conventional test statistic for first-order serial correlation used by econometricians.

The most important variable in explaining NG use per customer is found to be the heating requirements (HR) variable. Magnitudes estimated for this coefficient indicate that the percentage change in use per customer from a 10 percent change in this variable is estimated to be as great as 7.3 percent in New Mexico and to be as small as 4.7 percent in Mississippi. The estimated effects of the price of NG on NG use per customer, the NG (own-price) elasticities, while similar, are not the same in all states and are neither equal to zero nor to one.

The estimated electricity price effects (cross-price elasticity) are found to be statistically different from zero only in New Mexico and Arkansas (see Appendix I). The estimated elasticity is essentially the same in both state equations. Thus, any competition between electricity and NG within NG using households based on the price of electricity is small in two of the five states and nonexistent in the remaining three. The value for the Durbin-Watson (DW) statistic reveals that the relationship between months in the value of residuals, the unexplained variability in the gas use per customer, is not strong. In many instances the identification of a strong relationship between monthly values of residuals indicates that there is an important explanatory variable left out of the estimated relationship. For example, if the amount of time spent in the household was systematically changing over time, affected NG use per customer, and was independent of the other explanatory variables, a significant DW statistic would be expected.

REINVESTIGATION OF SOME KEY RESULTS FROM A PREVIOUS ECONOMETRIC STUDY

In their previously cited paper, Balestra and Nerlove[8] provided an econometric framework for investigations of RNG consumption using annual, state level data. Two major features of their framework were the use of random effects to represent the unexplained variability in demand across states and the use of NG demand in the previous period to explain NG demand in the current period.

In 1967 Balestra published a book[9] in which he reported estimated NG demand relationships by state. The overall objective of the study was to identify how economic factors affect the level of NG demand. The analysis of changes in the level of NG demand between years required an analysis of the effect of changes both in the number of customers and in the level of NG demand per customer between years.

Since 1966 most econometricians who have examined NG demand behavior have used Balestra and Nerlove's general model, except they have specified more general random effects, have used an additional factor or two, and have tended to limit analyses to the examination of demand per customer or per capita.

Balestra encountered many estimation and data problems in his investigations. For example, some of the data available to Balestra were very unreliable and several proxy variables were used in the final equations he estimated in his book. The very first equation estimated by Balestra in his book was:

$$G/P = a + bPG + cINC,$$

where,

G/P	= NG demand per capita by state,
PG	= average price of NG deflated by the consumer price index,
INC	= income per capita deflated by the consumer price index,
a,b,c	= estimated coefficients.

This equation was estimated by ordinary least squares by state for the time period 1950-1962. Balestra's results were uneven. He estimated positive coefficients for income, as expected, because an increase in income allowed consumers to substitute NG space heating units for wood-burning and coal-burning units during the 1950's and because income served as a proxy variable for all positive influences on the increase in the proportion of customers using NG for space heating during the period. However, Balestra also estimated many positive coefficients for price. This prompted Balestra to conclude: "The results obtained in the previous section point out that traditional demand analysis fails to give a satisfactory explanation of gas demand and suggests that a more dynamic approach is needed."

The more dynamic approach involved the use of a lagged dependent variable (the value of gas demand per customer in the previous year) to explain gas demand per customer in the current year. Balestra assumed that by using a lagged dependent variable he would be able to control for the relative fixity of the equipment stock. However, this created several problems for the estimation. In particular, the equipment stock was not relatively fixed at all in most states during the time period considered by Balestra. The stock of space-heating equipment of the average customer was increasing dramatically between 1950 and 1960 in many states. As a consequence, NG demand per customer in a current period was always greater than NG demand in the preceeding period. Accordingly, Balestra estimated coefficients for the lagged dependent variable that were greater than one in many instances. Moreover, within Balestra's theoretical framework, values greater than one for the lagged dependent variable implied negative depreciation rates. Naturally, Balestra's formal model assumed constant and positive depreciation rates for NG equipment used in the household.

In addition to interpretation problems, several statistical problems, such as biased and unstable coefficients for the other variables, also followed from Balestra's use of a lagged dependent variable.[10]

Balestra proceeded to estimate relationships for total NG demand rather than NG demand per customer. In the specification of relationships for total NG demand, moreover, Balestra assumed that the effect of the NG price on RNG demand per customer was very small or equal to zero. Many analysts after Balestra continued to make this questionable assumption because Balestra's text was generally considered to be the authoritative reference. When Balestra's equations for gas demand per customer are reestimated for the time period 1961 to 1981 for the five states examined here, different results from those obtained by Balestra are estimated. In particular, price is statistically significant (see Table 8.7). However, in this study the influence of equipment ownership on gas demand per customer is controlled by choosing states for analysis in which these equipment stocks, in particular the space heating equipment stock, are relatively fixed.

Table 8.7

A Comparison of Estimated Price Effects

| | Balestra[a] | | | | This Study[b] | | | |
| | Ordinary least squares | | | | Generalized least squares[c] | | | |
State	Price	t-stat	R^2	DW		Price	t-stat	R^2	DW
AK	+.08	+.22	.94	1.52		-1.00*	-4.05	.91	1.80
NM	+.52	+.46	.56	0.68		-.99*	-5.45	.94	2.03
MS	-.33	-1.19	.89	1.84		-.44*	-3.87	.81	1.65
OK	+.07	+.28	.82	1.74		-.67*	3.37	.72	1.72
LA	-.84	-1.61	.85	2.09		-.32*	-2.70	.74	1.90

[a] The income coefficient was found to be significant in all instances.

[b] The income coefficient was found to be insignificant in all instances.

[c] A generalized least squares estimation was used in which an estimated first-order serial correlation coefficient was used as a weight in the estimation. In both the original least squares estimation and in the generalized least squares estimation the coefficients were significant in all instances and the difference in the magnitude of the same coefficient for both estimations was less than 10%. An asterisk (*) indicates that a coefficient was statistically different from zero at the 5% significance level.

ESTIMATED PRICE EFFECTS FROM PREVIOUS STUDIES

A variety of own-price elasticities from previous studies were also examined. Most estimated short-term (between years) own-price elasticities reported in previous studies were relatively low in value (see Appendix II). Such results were consistent with Balestra's belief that these elasticities were near zero in value. However, these equations were estimated from combined data sets of state, annual data which equations included a lagged dependent variable.

A common procedure in estimations that include a lagged dependent variable is to divide the price coefficient (a) by one less the coefficient of the lagged dependent variable (b). The estimated magnitude is designated a long-run price elasticity. This estimated magnitude (a/(1 - b)) is almost always large and elastic because the denominator (1 - b) is almost always very small, since b is almost always near 1 in value. Thus, the importance of price is improved by making reference to the effect of price in the longer term. When equations are estimated (see Appendix II) based on the general economic model of Balestra and Nerlove (see Appendix III) which model, however, does not require a lagged dependent variable, results different from those in previous studies are obtained. The estimated own-price elasticities are generally found to be much nearer -0.5 than zero in value and statistically different from zero.

The absolute magnitude of the own-price elasticity is also understated in most previous econometric studies because the influence of measurement error is not taken account of in the estimation.[11] When the possible influence of the random measurement error is included as part of the estimation, the estimated own-price elasticities are again nearer -0.5 in value rather than zero. Moreover, significant negative own-price elasticities with either RNG demand per capita or per customer as the dependent variable have been estimated in most studies (see Appendix II). From these results alone it is clear that the own-price elasticity is not equal to zero as Balestra assumed.[12]

The estimated elasticities, besides being larger in absolute magnitude, are also easier to interpret when the elasticity is estimated with a model that does not include a lagged dependent variable. The estimated elasticities represent the average annual response of NG demand per customer to the price of NG during the time period. It is not necessary to combine the estimates of the own-price elasticity with the coefficient of the lagged dependent variable to obtain estimates of long-term price effects. Long-term price effects are merely the sum of a series of short term (annual) price movements. Estimates of the effect of changes in price on gains and losses in consumer surplus are thus easier to obtain[13]. Gains in consumer surplus from cost reductions are greater the larger the absolute value of the elasticity.

THE APPLICATION OF ELASTICITIES TO INDUSTRY BEHAVIOR

Some knowledge of the range of likely values for the own-price elasticity can be used to obtain a better understanding of why certain problems developed in the NG industry. One example of such a problem is the surplus underground NG storage capacity and the surplus NG supply that developed in the industry during the 1980's.

During the 1970's it was standard practice in the gas industry to withdraw NG from producing properties in the south-central United States during the late spring, summer, and early fall and to transport this NG to storage sites in the North. The NG was then continuously injected into storage sites until late fall. The industry in the northern United States needed to have sufficient amounts of NG stored underground at the close of the fall season to serve NG customers uninterruptedly during the winter months, especially RNG customers, as demands for space heating rose dramatically with the decline in the temperature. In general, when the overall level of RNG consumption increased, it was necessary to increase underground storage capacity.

In the early 1970's the real price of RNG began to rise. The rise in the price of NG during the 1970's reduced demand for NG per customer and, other things being equal, it should have reduced previously planned additions to NG storage capacity. However, the industry was still relying on forecasts which showed RNG demand continuing to rise due in part to the belief that the price of NG had little or no effect on NG use per customer. Since increases in real price during the 1970's were initially small, the effect of these increases in price on NG use per customer was not noticed. Then, between 1976 and 1977, a large increase in price occurred. However, the negative influence of price on NG use per customer was swamped by the positive influence of the very cold fall and winter of 1976. Thus, the influence of price on NG use per customer was still unrecognized. Other facets of the NG industry also supported the NG industry in its continuing to invest in underground NG storage.

There is a natural tendency for regulated industries such as the gas industry to overinvest. As regulated monopolies, the amount of income they earn is constrained by rate of return regulation. However, since earned income is the product of a regulated rate of return and a rate base, regulated firms can still maximize their income by maximizing their rate base, which includes investments such as NG underground storage even if the rate of return is fixed.

Overinvestment in underground storage had consequences for consumers because the capital cost of storage was passed on to them in the price they paid for NG. Thus, if there was more NG in underground storage than was necessary to adequately serve customer needs even in the coldest winters, consumers paid a higher price for NG without receiving any additional benefit. The cost of underground storage for each customer increased as underground storage capacity utilization declined. Moreover, each additional increase in the cost of NG reduced further the amount of NG each customer demanded, on average, which further increased excess capacity and the cost of storage for the average customer. Some relief for existing customers from this situation occurred when sufficient numbers of new customers were added to the distribution system; the fixed cost was then spread over these additional customers. An indication of the capability of the NG industry to serve weather-sensitive customers can be obtained by examining the ratio of total storage capacity relative to residential and commercial consumption of NG over time (see Table 8.8). An examination of this ratio for the years 1972, 1977 and 1988 depicts well the excess underground storage that had developed in the United States by the early 1980's.

Even though 1972 was the peak year for aggregate RNG consumption and 1976 was the peak winter demand year, storage continued to be developed.[14] During the 1970's, curtailments also occurred in the heating season (November-March) of 1976/1977, in part, because underground storage reservoirs were not filled to planned levels, large withdrawals occurred even before the onset of the heating season because of unseasonably cold weather, and wells froze in the South, which affected deliverability. The impact of the fall and winter of 1976/1977 on the natural gas delivery system also spurred some additional investment in underground storage. However, the values for the RATIO statistic indicates that there was a large increase in ultimate capacity relative to the aggregate level of

residential and commercial consumption in all areas of the country except for the
Mid-Atlantic region.

Table 8.8

Natural Gas Underground Storage Capacity as a Proportion of Total Deliveries of Natural
Gas to Residential and Commercial Customers

Division	1972	1977	1982
New England	0	0	0
Middle Atlantic	0.78	0.88	0.81
East North Central	1.02	1.26	1.27
West North Central	0.54	0.75	0.91
South Atlantic	0.87	0.93	1.02
East South Central	0.81	0.87	1.06
West South Central	1.04	1.47	2.00
Mountain	0.71	0.89	1.19
Pacific	0.39	0.61	0.59
United States	0.78	0.99	1.06

Sources: Ultimate Reservoir Capacity, American Gas Association, *Gas Facts*, various
years; Deliveries, Energy Information Administration, *Natural Gas Annual 1984*, Vol. 1,
DOE/EIA-0131 (Washington, D.C.: Energy Information Administration, 1985).
   Note: For Census Division definitions, see **Table 6.6**.

The availability of more storage capacity than was necessary to satisfy customer
demands was indicative of the overall surplus of NG that was developing in the
NG industry during the time period and which was acknowledged by the NG
industry in 1982 when some started to use the term "gas bubble" to characterize
the situation.[15] At then-current prices, producers and pipelines were willing to
produce much more NG than they were able to sell. The increased cost associated
with increased storage and other capital expenditures was coupled with increased
costs associated with newly purchased gas by pipeline companies.[16]
   Now, a common perception within the gas industry during the late 1970's and
1980's was that any additional amounts of NG that could be obtained could be
sold. New customers would be added, and each customer would continue to use
about the same amount of NG per appliance. The notion was that there was this
great pent-up or "latent" demand for NG. This bright vision for many in the NG
industry may have begun to fade after 1979. Between 1979 and 1980 RNG
consumption per customer declined noticeably - from 115 Mcf to 108 Mcf, or 6
percent. This decline probably surprised many managers in the industry. Although
heating degree days (weighted by spaceheating customers), declined by 2.3 percent
between 1979 and 1980, which is consistent with a decline in demand, two other

important changes occurred which are consistent with an increase in demand. The residential price of electricity in constant dollars increased by 5.9 percent and the number of space-heating customers relative to the number of RNG customers (saturation) increased by 2.6 percent. Since space-heating customers during the 1970's consumed about three times as much gas as otherwise similar customers, this relatively large increase in saturation should have increased use per customer by about 7.8 percent.

Some indication of the effect on expected revenues received in 1980 from existing customers as of 1979 from the 21 percent increase in the real price of NG during 1980 can be obtained by employing the following relatively simple equation:

$$(GD_{1979} \times E \times dGP) \times CUST_{1979} \times GP_{1980}$$

where,

GD_{1979} = average NG demand per customer in 1979,

E = estimated own price elasticity,

GP_{1980} = average residential price for NG in 1980,

$CUST_{1979}$ = average number of residential customers in 1979,

dGP = proportionate change in the real price of RNG between 1979 and 1980.

Using the above equation, the estimated reduction in demand per customer and the reduction in revenues resulting from the 21 percent increase in the real price of NG between 1979 and 1980 is reported in Table 8.9. Now, if we merely assume that the combined effect of the increase in the price of electricity and of saturation and of the decrease in heating degree days cancelled out to zero, and if other factors did not have much of an effect on NG use per customer, then the 6 percent decrease in use per customer and the 21 percent increase in price is consistent with a value for the own-price elasticity of about -0.3. This value for the elasticity is near the lower bound estimate for the previously discussed elasticity when the possible influence of measurement error is counted in the estimation.

This value would have produced a reduction from expected real revenues equal to approximately $1.17 billion for the gas industry, conducting business under the belief that the elasticity was equal to zero. If the elasticity were equal to -0.5, which is our best estimate, then the reduction in expected revenues would be 1.94 billion.

The significance of the results in Table 8.9 can be gauged better by recognizing that total revenues received from RNG customers in 1979 were 14.77 billion

Table 8.9

Estimated Reduction in Natural Gas Demand per Residential Customer

Elasticity	Reduction in demand per customer in 1000Mcf	Reduction in revenue in million dollars
-.7	-16.90	-2,720
-.6	-14.49	-2,405
-.5	-12.07	-1,943
-.4	-9.66	-1,555
-.3	-7.24	-1,165
-.2	-4.83	-777
-.1	-2.41	-387
0	0	0

dollars. Thus, the estimated reduction in revenues of $1.17 or $1.94 billion between 1979 and 1980 represented 8 percent or 13 percent less, respectively, than the revenues that the industry may have been expecting to receive from residential customers during that year. Thus, the industry's entering into high cost contracts for the purchase of NG is explained by the industry's expectation that revenues would be much higher than the revenues actually received.

In general, different assumptions about price effects lead to different conclusions about the impact on residential consumers of private and public sector policies. Assumptions about price effects have also influenced both private and public policy decisions in the past and will continue to influence such decisions in the future. The lack of interest of the public sector during the 1980's in promoting policies that reduce the cost of RNG stems in part from the erroneous belief that the effect of price on use per residential customer is negligable. As a consequence the public sector continues to underestimate the increase in social welfare from policies to reduce the cost of NG.

NOTES

1. W. H. Babcock, S. B. Siegel and C. Swanson, *Energy Demands 1972-2000*, PRC-D-2017 (McLean, Va.: PRC Energy Analysis Company, May 1978). Energy Information Administration, *The Natural Gas Market Through 1990*, DOE/EIA0366 (Washington, D.C.: Energy Information Administration, May 1983).

2. P. Balestra, *The Demand for Natural Gas in the United States* (Amsterdam: North Holland Publishing Company, 1967).

3. P. Balestra, and M. Nerlove, "Pooling Cross Section and Time Series Data in the Estimation of a Dynamic Model: The Demand for Natural Gas," *Econometrica*, 34, (1966): 685-613.

4. L. A. Doman, J. H. Herbert and R. Miller, *An Assessment of the Quality of Selected EIA Data Series - Energy Consumption Data* (Washington, D. C.: Energy Information Administration, 1986). J. H. Herbert, "Measurement Error and the Estimation of Regression Equations - A Case Study," *Proceedings of the Section on Economic and Business Statistics*, 1987 Annual Meetings of the American Statistical Association, (Alexandria, Va.: American Statistical Association, 1988) 187-190.

5. Income per household is not included initially as an explanatory variable primarily because most previous econometric investigations have found an estimated income coefficient not to be significantly different from zero in equations in which the dependent variable was NG use per customer. An exception was the early work of Balestra in which income was a proxy variable for the growth in the number of customers that used NG for space heating. Income per customer affects fuel choice in terms of the movement from inferior sources of energy to superior sources of energy but at the level of the individual household changes in the level of income per household or per customer between years do not much influence the level of use. Results from including an income variable in the estimated equation, however, are included in Appendix I.

6. Temporal observations on heating degree days are not examined because the relationship between heating degree days and NG demand is very well established. When heating degree days increase there is an increase in demand. J. H. Herbert, "Data Analysis of Sales of Natural Gas to Households in the United States," *Journal of Applied Statistics*, 13 (1986): 199-211. J. H. Herbert, and E. Kreil, "Specifying and evaluating aggregate monthly natural-gas demand by households," *Applied Economics*, 21 (1989): 1369-1381.

7. The fact that the proportion of customers that were space-heating customers was relatively constant is convenient for analytic purposes because available data on changes in this proportion between years are frequently very poorly measured.

8. Balestra and Nerlove, "Pooling Cross Section and Time Series Data," 685-613.

9. Balestra, *The Demand for Natural Gas*.

10. The use of a lagged dependent variable in RNG demand relationships can also cause severe multicollinearity problems as demonstrated in J. H. Herbert and K. Dinh, "Reporting the Uncertainty in Regression Coefficients from Errors-in-Variables," *Proceedings of the Section on Economic and Business Statistics*, 1988 Annual Meetings of the American Statistical Association (Alexandria, Va.: American Statistical Association, 1989), 259-264. J. H. Herbert and L. Barber, "Regional Residential Natural Gas Demand, Some Comments," *Resources and Energy* (1988): 387-391. Moreover, the use of lagged dependent variables may mask specification problems which are difficult to identify with

conventional regression diagnostics. T. Norstrom, "Lagged Response or Omitted Predictor: The Case of Government Popularity," *Social Science Research*, 16 (1987): 119-137.

11. Diagnostic formulae for evaluating the bias in estimated coefficients from random measurement error as applied to NG demand relationships can be found in J. H. Herbert and P. S. Kott, "An Empirical Note on Regressions with and without a Poorly Measured Variable Using Gas Demand as a Case Study," *The Statistician*, 37 (1988): 293-298. Non-standard formulae for systematic measurement error are derived in J. H. Herbert and K. T. Dinh, "A Note on Bias from Proxy Variables with Systematic Errors," *Economics Letters* (1989): 207-209. These non standard formulae are applied in J. H. Herbert, and K. Dinh, "An Empirical Note Concerning Systematic Measurement Errors in Regression Analysis," *Journal of Applied Statistics* (1990): 435-441. Particular types of measurement error have been shown to exist for all the explanatory variables used in RNG demand relationships. In the commonly reported least squares estimates only a random error either in the equation, which included NG use per customer on the left hand side of the equals sign, or in NG use per customer is allowed for in estimating the coefficient. In general, both the theory on bounds for estimated regression coefficients and more conventional asymptotic formulae for bias indicate that the least squares solution will be attenuated or will form the lower bound.

12. Another generalization that follows from an examination of the magnitude of the estimated elasticities is that the elasticities generally increase when the regression equation represents consumers in more northern climes. Thus, it is understandable why northern constituencies of congressional representatives, especially where a large proportion of the households are using NG, have strongly supported policies that would result in a reduction in the cost of NG. Reductions in the cost of NG apparently result in larger gains for northern customers than for southern customers, that is for each percentage decrease in the cost of NG, the percentage increase in the amount of NG service RNG customers are willing to purchase is larger for northern customers than for southern customers.

13. M. G. Baumann and J.P. Kalt, "Intertemporal Consumer Surplus in Lagged Adjustment Demand Models," *Energy Economics* (January 1986): 2-11.

14. For a more extensive discussion of natural gas storage see J. H. Herbert and Jim Tobin, "Recent Trends and Regional Variability in Underground Storage Activity," *Natural Gas Monthly February 1991* (Washington, D.C.: Energy Information Administration, 1991).

15. For a discussion of the surplus NG that developed in the 1980's and the implications of this surplus see J. Heinkel, M. Carlson, J. H. Herbert, B. Mariner-Volpe, D. Morehouse, J. Tobin, W. Trapmann, and D. Van Wagener, *Assessment of Pipeline Capacity to Transport Domestic Natural Gas Supplies to the Northeast*, (Washington, D. C.: Energy Information Administration, 1989). For some examples of the consequences of the excess NG capacity on the NG industry see J. Tobin, D. Helme, and J. H. Herbert, *Growth in Unbundled Natural Gas Transportation Services: 1982-1987* (Washington, D. C.: Energy Information Administration, 1989). For examples of the surplus at the site of production site see P. Shambaugh, J. Tobin, D. Van Wagener, and J. H. Herbert, *Natural Gas Production Responses to a Changing Market Environment 1978-1988 (Washington, D.C.: Energy Information Administration, 1990).*

16. The terms of these contracts required pipeline companies to purchase NG even if they could not sell it. These terms were known as take-or-pay terms. During the early 1980's

several pipeline companies found themselves in a situation where they were not able to market much of the new highpriced NG they had contracted for in take-or-pay contracts. This situation presented the companies with difficult problems which even threatened their existence. This is much different from the 1960s when it was claimed that the regulated price was not great enough to attract sufficient supplies of NG for residential customers, or that sufficient additional discoveries were not being discovered at prevailing prices to serve customers with a secure supply of NG in reserve, and that these conditions increased uncertainty within the NG industry, which during the 1970's resulted in restrictions on adding new customers. P. W. Mac Avoy and R. S. Pindyck, *The Economics of the Natural Gas Shortage* (New York: North Holland, 1975).

9

SUMMARY AND CONCLUSION

NG delivery to households has evolved during the twentieth century from being privately distributed by many farmers to being distributed by some of the largest companies in the world. By 1970 Pacific Gas and Electric Company had more RNG customers than the entire country had in 1910. Yet even in 1910 in parts of the country near NG-producing sites, the average customer used more of this clean and convenient fuel within the household than today's customer uses in California. Customers in Pennsylvania, West Virginia, New York, and Ohio used NG as extensively in the household in 1910 as in 1970 when it was similarly inexpensive (see Table 9.1).

Table 9.1

Natural Gas Use per Customer and Price in 1910 and in 1970

	Use per Customer[a]		Constant 1970 Price[b]	
	1910	1970	1910	1970
Pennsylvania	135	134	1.04	1.24
West Virginia	157	166	0.72	0.89
New York	115	90	1.23	1.40
Ohio	139	188	1.05	0.90
United States	128	125	0.95	1.09

Sources: Use per Customer and Price, *Mineral Resources*, 1910, *Natural Gas Annual 1988*, Table 23; Deflator, Consumer Price Index - All Items, *Historical Statistics, Census*, 211.

[a] In thousands of cubic feet.

[b] In deflated dollars per thousand cubic feet.

The dangers inherent in gas use were also much less in 1970 than in 1910. This occurred because the technology for gas distribution within and outside the household had greatly improved, because government got involved in the industry in intervening years, and because newspapers and consumer-oriented magazines constantly reported the explosions and poisonings that resulted from careless use of this convenient but dangerous fuel.

Today, the current and future conditions of NG markets are monitored closely by a well-integrated industry. Under ordinary conditions, the available reserves of NG are evaluated annually by industry groups and by agencies of state and national governments. Under extraordinary conditions, such as the cold December of 1989, the capability of the NG system to redistribute NG on short notice is monitored hourly by operators of pipelines and underground storage systems. Summary reports of the condition of the market are reported to the industry and government agencies in the trade press on a daily and weekly basis. Things were much different near the turn of the century when through a combination of businesses operating independently and of government indifference some markets failed in particular years when residential customers were left without gas during the coldest part of the winter. Things became so bad in Indiana that many markets didn't just fail in particular years but they collapsed for longer periods. The number of RNG customers in Indiana consistently declined from 102,000 in 1902 to 27,000 a decade later.

Over the years large improvements have been made not only in the ability to deliver NG to markets but also in the efficiency of NG use. Moreover, these efficiency improvements have frequently resulted in health benefits as well. In the second decade of this century, approximately 16 percent of the volume of gas delivered to a residential customer may have been wasted through leakage[1] within the household which caused headaches or worse if enough gas collected in the house. Other improvements in efficiency saved the energy of the homemaker as well. Many burners during the 1920's were set at 3 instead of 2 ounces pressure and were not adjusted to use the proper gas and oxygen mixture which caused the gas to burn with a 6 1/2-inch high yellow flame instead of a 4-inch high blue flame, scorching the sides of pots, taking 15 percent more gas to boil a pot of water, and taking the homemaker several minutes to scrub each burned pot.[2]

Gas lighting was very popular near the turn of the century and was still popular in 1920, when 65 percent of American homes used gas for lighting. Many of these households used environmentally damaging, expensive MG instead of the environmentally benign, relatively inexpensive NG.

During the 1920's the allocation of NG within the United States changed dramatically. Parts of the industrial sector supported by cheap NG were growing phenomenally. The residential sector was paying a constantly increasing price for NG. More NG was used by industrial customers in Texas in 1929 than was consumed by all customers in the United States in 1910. Some industrial firms, such as carbon black and natural-gas gasoline plants, were able to relocate to Texas to take advantage of the relatively inexpensive high pressure NG there. However, residential customers in the North would not be able to take advantage of this NG because the NG industry had not yet completed several tasks: completing the many mergers and combinations begun during the 1920's, beginning to invest in underground NG storage, and deciding which pipeline systems would serve particular markets with long-distance pipelines. By 1929 the residential sector was receiving only one-fifth of the NG produced in the United States; in 1919 it had been receiving two-fifths. Nonetheless, the volume of NG delivered to residential customers increased by 40 percent between 1919 and 1929.

As major industrial markets for NG were developed in Texas and California during the 1920's, major residential markets were established there as well. However, the residential customer in Texas and California paid, on average, seven to eight times as much for NG as the industrial customer, and residential customers in these states paid as much for NG as residential customers in states that did not produce any NG.

Residential customers in the United States paid much more for NG during the 1920's than in any previous decade. Thus each customer was consuming much less NG. Despite such changes in society as increased urbanization, which were favorable to the development of RNG markets, there was little growth during the 1920's because the industry did not aggressively pursue the development of new markets in the North.

Refrigerators and improved water heaters had been developed by the 1920's. But because of technical problems, few were sold. NG air conditioners were in the development stage. Yet within the households, only fans were available for cooling; ice boxes were used for refrigeration; and water was frequently heated on top of the stove or with an attachment on the furnace. Separate water heaters serving all the hot water needs of the household were not common. A wide variety of ornate, durable, iron stoves and space heaters were placed in positions of distinction within many households. The stoves were not to be replaced completely by colorful, more technically advanced, yet nearly indistinguishable stoves for several decades. Practically invisible central space units appeared only in the households of the wealthy. Since coal and wood were widely used as sources of household energy, the increased availablity of NG meant a much cleaner environment within the household and much less expenditure of energy by household members to support the needs of the household for space heat, clean clothes, and other services requiring energy.

NG was increasingly used during the 1920's for gasoline production to power cars and trucks and for electricity production to power vacuum cleaners and other new electrical household appliances. Because of the slow growth in RNG markets, this growth probably improved the economic welfare of consumers as much as its increased use in households.[3]

Although reports on the NG industry at the time suggest that a general exhaustion of NG wells was taking place during the 1920's and the 1930's in the northern region, examination of the complete historical record does not bear this out (see Table 9.2). The amount of NG produced and marketed (marketed production) in the Appalachian region (the northern region plus Kentucky) was relatively constant during the 1920's, and fell to its lowest levels ever during the early 1930's in response to the general economic decline. It rose to very high levels during the two world wars, when government intervened in the management of the industry. During the 1980's government again intervened with similar results. By allowing producers to receive a high, regulated price for particular categories of wells, such as devonian shale and low production (stripper) wells, common in the region, production expanded. Thus, the long-term production profile in the northern region indicated a highly flexible production capability.

TABLE 9.2

Eighty Years of Marketed Production of Natural Gas in the Appalachian Region (in billion cubic feet)

	Year									
	0	1	2	3	4	5	6	7	8	9
1910s	373	372	418	424	425	446	510	522	461	420
1920s	437	320	360	389	354	342	352	337	341	363
1930s	334	292	251	250	285	308	352	385	332	367
1940s	385	422	444	471	430	386	409	496	432	368
1950s	401	438	398	402	442	423	412	408	407	415
1960s	439	424	412	418	403	410	425	436	461	447
1970s	453	465	446	448	454	392	406	415	444	445
1980s	466	502	478	464	583	582	588	586	613	643

Sources: *Mineral Resources of the United States*, 1909-1923, *Minerals Yearbook*, 1924-1931, *Minerals Yearbook*, 1932-77, *Natural Gas Annual*, 1978-1989.
Note: The Appalachian region includes the states of West Virginia, Ohio, Pennsylvania, New York, and Kentucky.

At the close of the 1920's, the economic and environmental advantages of NG service were clearly acknowldeged in and around Atlanta, when Southern Natural Gas completed a pipeline from Louisiana to Georgia. This pipeline was one of several long-distance pipelines that were completed near the beginning of the Great Depression.

As incomes declined precipitously and as prices continued to rise during the Great Depression, NG consumption per customer declined as well. Average RNG use per customer fell to levels similar to those observed in the Indiana market after the failure of that market near the turn of the century. The size of the reduction suggests that consumers reduced the amount of NG they used for space heating and other end use services for NG in the household. Available evidence, moreover, suggests that consumers did not switch to electrical energy, because electrical energy was several times as expensive as NG, nor did they switch to fuel oil, because of the conversion cost involved and because the cost of fuel oil on a Btu eqivalent basis was about equally expensive. Instead, they switched to inferior sources of energy such as wood and coal, which they burned in fireplaces and in inexpensive room space-heating units, and they simply conserved on their use of energy services in the households. These changes resulted in a significant decline in the standard of living within many households in the United States.

Even when income began to increase again in the 1930's, NG consumption per customer remained relatively constant. As a consequence, NG distribution companies began to market NG furnaces and appliances in earnest. They also provided information about conservation in an attempt to attract more customers and to motivate existing customers to add new NG appliances to their stock of appliances. The industry also began to extend distribution channels into regions where gas was manufactured from coal and other hydrocarbons because NG was not to be found nearby.

Possibly due to the stagnant growth in sales of NG to domestic customers in the 1920's and in the 1930's in the northern region, the industry began to put underground NG storage in place. This enabled an increasing amount of relatively inexpensive NG to be imported from the South to the North. This tended to reduce the average cost of NG and also to improve wintertime deliverability capability in particular areas of the North. Thus, a major technological constraint to the further development of RNG markets began to be overcome at the close of the 1930's.

During the 1930's the iron-fisted control of large NG pipeline companies over market development was well documented in several government publications. Markets were effectively closed to new entrants through a variety of devices. Because of this control and other abuses of power by the NG industry, consumer groups were formed. These groups attempted to by-pass the local gas utility and pipeline and to make arrangements for the purchase of NG from new suppliers of NG in the south-central United States who were willing to sell NG at a relatively low cost because of their access to large amounts of high pressure gas. This type of arrangement resurfaced in the 1980's and was called "by-pass." The difference between the 1930's and the 1980's is that in the 1980's it was mostly large commercial and industrial customers attempting to reap the benefits of bypass and not groups of residential customers. The principal consumer group in the 1930's, the Cities Alliance, wanted to establish a spot market for NG in which the average price of NG was more a consequence of supply and demand conditions rather than of stipulations in long-term contracts. The purpose of many of these stipulations was to protect the NG industry from changes in economic conditions when they were unfavorable to the industry.

Many recommendations for broad governmental involvement in the NG industry can be found in statements made at the hearings for the Natural Gas Act (NGA) of 1938 and in reports published as analytic support for these hearings. Yet the major role for the government according to the NGA was to be a protector of the value of the capital equipment of the NG industry. Despite technical advances in some areas, safety continued to be a major problem. The dangers inherent in gas use to domestic customers was demonstrated dramatically in 1937. In that year 244 people were blown up in a school in New London, Connecticutt.

At the beginning of the Second World War the major sources of energy for the household were still coal and wood. The accelerated transition from households dependent on wood, coal, and manual labor for energy to households dependent on NG, electrical energy, and fuel oil would have to wait until the close of the Second World War as increasing amounts of NG were allocated to the war effort. Large amounts of NG were being used to produce high-octane gasoline and carbon black for the war in Europe. The national government helped in making management decisions for the NG industry and funded the production of pipeline systems. Government involvement in the industry, huge new NG reserves, income growth, NG price reductions in constant dollars, low interest rates, and social and demographic changes during and after the Second World War would result in accelerated growth once the war was over.

The competition between NG and electricity, which had begun with the lighting market near the turn of the century, broadened after the Second World War to include competition for the refrigeration and air-conditioning markets. Although the cost of operation was less for NG appliances than for electrical appliances, and NG units were quieter, electrical units had the distinct advantage of being less expensive and of being technologically superior in part because the electric industry invested more in research.

Air conditioners, clothes dryers and other appliances powered by NG and electrical energy began to allow the household economy to operate independently of the vagaries of the weather for the first time. Thus, the availability of appliances and the affordability of energy after the Second World War began to affect the organization of social and economic life within the household.

The growth in the number of residential customers and in use per customer by 1950 indicated that RNG markets in the north for the first time could not possibly be supplied with indigenous supplies of NG. Thus consumers in the North would become increasingly dependent on energy from the South if they were to improve their quality of life by increased use of relatively inexpensive, flexible, and clean NG.

As both investments in underground NG storage and in pipeline systems throughout the United States were increased during the 1950's and 1960's, the number of residential NG customers grew dramatically. NG use per customer began to rise once more to levels observed during the second decade of the twentieth century. Only the enormous fleet of woefully inefficient gasoline powered automobiles, and the unavailability of NG along the eastern seaboard, where fuel oil was increasingly being used, kept NG behind petroleum as the primary source of energy in the American economy during the great growth period for NG which ended in 1972, a year before the Arab oil embargo.

The northern states were still the major markets for RNG in the United States in the 1970's. Of the thirteen states in 1972 having sales of NG to residential customers greater than 100 Bcf, nine were in the north. The remaining four states were California, Texas, Kansas, and Missouri (with sales just over 100 Bcf). These markets had developed under generally different conditions than more northern and eastern markets.

Texas had huge supplies of NG and minimal requirements for NG storage to adequately serve domestic customers. California initially had huge supplies of NG, several companies aggressively seeking market development, and minimal requirements for NG storage. Important RNG markets developed early within major cities in eastern Kansas and western Missouri because these cities were readily accessible to major producing areas in the early part of the century. The development of the residential market in Missouri received a boost when the utility in St. Louis in the eastern portion of the state began receiving NG service on a regular basis from the Monroe field in Louisiana in the early 1930's by means of a long-distance pipeline. The number of RNG customers grew from 104,000 in 1930 to 377,000 in 1939. However, it was not until 1950 that the gas utility in St. Louis received enough NG to switch from mixed gas service to straight NG service. Between 1948 and 1955 RNG use per customer in Missouri increased from 61 Mcf to 124 Mcf as many customers began to use NG for space heating.

The Arab oil embargo and rising NG prices during the 1970's brought the growth in use per customer in Missouri and in most other states to a close. Use also declined because of better and more information about energy in the news media. A major recession in 1974/1975 reduced the growth in the number of customers as did NG utility restrictions on new space heating customers. Despite rising prices the NG industry was unwilling to supply NG for new customers. Deliverability problems during the extremely cold winter of 1976/1977 demonstrated the inability of the NG industry to adequately serve established markets. With this troubled setting as a background, the NGPA, which began in a very complicated way to decontrol NG prices, was passed in 1978. As a consequence of this legislation, RNG prices would increase.

Dramatic changes occurred in NG prices in the ten years after the passage of the NGPA. The price of NG rose for all customers between 1979 and 1983. Some have estimated that the cost to residential customers of the deregulation of NG under NGPA, with its many complicated categories of NG for pricing purposes, instead of total decontrol of NG prices, resulted in residential customers paying

$8.6 billion more while industrial consumers paid $69.9 billion less because of price discrimination.[4]

The large price increase after 1978 led to a surplus of NG in the industry because the industry had once more underestimated the effect of price increases on use per domestic customer and because the price increase was coupled with plummeting oil prices in 1981 and a severe recession in 1982. However, between 1983 and 1988, the average price of NG to residential customers fell from $6.08 to $5.47 per Mcf. The price of NG to industrial customers, moreover, fell by much more because of special programs initiated by the Federal Energy Regulatory Commission, which attempted to reduce the surplus of NG. By 1988 many industrial customers were able to obtain NG at prices between $1.30 and $2.30 on spot markets while the average domestic customer was paying three to four times as much for NG.

NG use per customer was observed to decline as price rose during the 1970's. Yet few anticipated the magnitude of the decline since the effect of price on use per customer was seriously underestimated by many. As a consequence, public representatives and the gas industry pursued questionable policy objectives. Because they underestimated the capability of NG customers to reduce their use of NG when price rose, consumer advocates consistently overestimated the reduction in economic welfare from these increases. Moreover, policy analysts in the private and public sector consistently used very unreliable estimates of price effects in their forecasting equations. The end result was that the participants in the NG debate in the United States were constantly surprised by real world outcomes and frequently supported errant policies.

As we proceed into the last decade of the twentieth century, NG use in the household appears to be at a crossroads with electrical energy. Some statistics suggest that a growing and larger percentage of new homes are equipped with electrical rather than with NG space-heating units. This appears to be part of a broad trend towards the further electrification of our society. On the other hand, recent Bureau of Census statistics show that in the latter part of the 1980s a much larger percentage of new homes were being equipped with NG space heating units rather than with electric units (see Table 9.3).

However, two applications for NG - automobiles and air conditioners - may steer the residential energy market clearly toward greater NG use when the environmental and economic implications of this move are fully understood. The increased use of NG powered automobiles, whether in the form of compressed natural gas (CNG) or methanol, will help to reduce the many societal maladies of the late twentieth century associated with hydrocarbon and other emissions from gasoline powered automobiles.

The greater use of CNG vehicles should result in reductions in imported oil and in the risk of international conflicts that a dependency on imported oil creates. CNG vehicles are also quieter. And in 1990, when the cost of unleaded gasoline was more than $1.25 for much of the year, the reduced operating cost of CNG vehicles is particularly attractive. The cost of energy from CNG is equivalent to gasoline at $0.80 a gallon.[5] Environmental advantages of CNG vehicles when compared to gasoline powered vehicles are significant. There are reductions of 80 percent or better in the amounts of reactive hydrocarbons and of carbon monoxide emissions and reductions of 25 percent or greater in the amount of nitrogen oxide and of carbon dioxide. Air toxics such as benzene, tuolene, and polycyclic aromatics are also reduced.[6]

Table 9.3

Percentage of New Houses with Different Types of Heating Systems

Heating system	1980	1981	1982	1983	1984	1985	1986	1987	1988	1989	1990
Gas	41	41	40	43	45	44	47	52	54	58	59
Electricity	50	50	50	49	48	49	44	40	37	34	33
Oil	3	2	3	2	2	3	5	5	6	5	5
Other/None	5	7	8	6	5	4	4	3	3	3	3

Source: U. S. Department of Commerce, Bureau of the Census and U. S. Department of Housing and Urban Development, *Characteristics of New Housing*, 1980-1990.

The negative features of CNG vehicles, however, are several.[7] It generally takes much longer to refuel a CNG vehicle than a gasoline powered vehicle. The retrofitting of an existing gasoline powered vehicle may cost several thousand dollars. Although a country such as Italy, which has more than 300,000 NG vehicles, have experience with the mass production of CNG vehicles, the cost and performance characteristics of mass-produced CNG vehicles in the United States is still an unknown. Nonetheless, CNG-powered vehicles drove off assembly lines in 1991.

Since NG air-conditioning units use ammonia instead of freon as a circulationg fluid, increased use of gas units instead of electrical units in buildings will result in a reduction in freon use, which is associated with damage to the ozone layer. The general use of NG air conditioners instead of electrical units in cities such as New York will also make less compelling the argument that unpopular nuclear reactors are necessary to generate electrical energy to satisfy summer air-conditioning demands.

Because of the seasonal nature of air-conditioning use in the United States, the increased use of this technology by residential customers will reduce the monthly variability in demand for NG by these customers. Since a reduction in the variability of NG use results in a reduction in the cost of NG service, the cost of

NG to residential customers will fall when the use of residential NG air conditioning increases.

Thus residential customers are once more in a position where they may reap direct economic and environmental benefits from increased NG use. If consumers and their representatives act forcefully, much as they did during the Great Depression, then a new era may begin. In this new era consumers will gain significant economic and environmental advantages similar to those gained in the 1930's when residential consumers switched from environmentally damaging and expensive MG to environmentally sound and less expensive NG.

NOTES

1. H. P. Westcott, *Handbook of Natural Gas* (Erie, Pennsylvania: Metric Metal Works, 1920), 530.

2. Idem., 539.

3. Interestingly enough, the future of the NG industry in 1989 hinges on the increased use of NG to power cars and to produce electrical energy. R. L. Itteilag, "Gas Demand Outlook," presented at the *11th Annual American Conference of the International Association of Energy Economics*, October 17, 1989.

4. R. C. Sickles, and M. L. Streitwieser, "The Structure of Technology, Substitution, and Productivity in the Interstate Natural Gas Transmission Industry Under the Natural Gas Policy Act of 1978," *Working Paper, Center for Economic Studies*, Bureau of the Census, Washington, D.C., September, 1989.

5. American Gas Association, "An Analysis of the Economic and Environmental Effects of Natural Gas as an Alternative Fuel," *Economic Analysis, EA 1989-10*, Dec. 15, 1989.

6. W.H. Kohl, ed., *Methanol as an Alternative Fuel Choice* (Washington, D.C.: Johns Hopkins Foreign Policy Institute, 1990).

7. "Alternatives to Oil Move from Lab to Road," *New York Times*, August 28, 1990.

Appendix I

ADDITIONAL ESTIMATIONS

In addition to regression results reported in Table 8.6, additional equations, which are reported below, were estimated. Since indications of first-order serial correlation were expected to be observed after fitting an equation, estimations with corrections for first-order correlation were automatically performed.

An income variable (INC), real income per capita, expressed in logarithmic form, was considered. This was done because it is conventional to include income in most demand relationships. An income variable was not included in the estimations reported in the main text because of the ambiguity associated with the relationship between use per customer and income. Increases in the number of customers has historically been related to income growth, as NG was substituted for inferior fuels after the Second World War. The dramatic income reductions that occurred at the time of the Great Depression affected use per customer, but in most published estimations, using historical data income per household or per capita, whether personal income or personal disposable income, has been found to be unimportant in explaining use per customer.

When the inferior sources of energy such as coal, wood, and in some instances fuel oil were still widely used, households that were hooked up to a gas line or able to be hooked up to a gas line substituted NG for the inferior fuel as their income grew. This occurred in the Northeast as late as the 1960's, when consumers substituted NG appliances for coal and fuel oil furnaces even though the cost of NG in equivalent energy units was greater than the cost of these fuels.

As their income continued to increase, however, some of these consumers, who considered electricity a safer and superior source of energy for the household, would substitute electric central space heaters, ranges, and other appliances for NG appliances. Thus, the relationship between NG use and income growth would be negative and not positive. Additional discussion about the ambiguity in this relationship can be found in J. H. Herbert, "Data Analysis, Specification, and

Estimation of an Aggregate Relationship for Sales of Natural Gas per Customer,"
Journal of Economic and Social Measurement, 6 (1986): 165-174.

State   Coefficients (Standard Errors in parenthesis)

	GP	HR	D1974	EP	INC	R^2	DW
1. AK	-.76(.089)	+.40(.162)	-.10(.047)			.81	.66
2. AK*	-.62(.214)	+.55(.098)	-.05(.024)			.94	1.92
3. AK	-.85(.050)	+.53(.094)	-.06(.027)	+.17(.028)		.94	1.51
4. AK*	-.82(.077)	+.52(.100)	-.05(.025)	+.16(.036)		.95	1.88
5. AK	-.75(.054)	+.57(.102)	-.06(.029)		-.19(.034)	.94	1.42
6. AK*	-.73(.070)	+.56(.111)	-.04(.025)		-.18(.046)	.94	1.76
7. AK	-.83(.077)	+.54(.102)	-.06(.028)	+.14(.101)	-.04(.117)	.94	1.52
8. AK*	-.81(.092)	+.53(.109)	-.05(.026)	+.14(.123)	-.03(.144)	.95	1.86
1. NM	-.48(.037)	+.75(.156)				.93	.77
2. NM*	-.49(.060)	+.71(.102)				.96	1.85
3. NM	-.47(.029)	+.75(.119)		+.14(.039)		.96	1.29
4. NM*	-.48(.043)	+.73(.106)		+.14(.054)		.97	1.79
5. NM	-.41(.065)	+.76(.140)			-.19(.063)	.96	1.32
6. NM*	-.38(.043)	+.78(.128)			-.16(.088)	.96	1.78
7. NM	-.59(.116)	+.70(.124)		+.34(.166)	+.28(.239)	.96	1.22
8. NM*	-.64(.131)	+.65(.119)		+.37(.179)	+.35(.260)	.97	1.81
1. OK	-.46(.065)	+.58(.131)	-.08(.041)			.79	2.31
2. OK*	-.45(.052)	+.65(.125)	-.07(.041)			.81	1.82
3. OK	-.46(.068)	+.57(.139)	-.09(.045)	-.09(.039)		.79	2.29
4. OK*	-.45(.059)	+.65(.138)	-.07(.049)	+.00(.031)		.81	1.81
5. OK	-.47(.067)	+.57(.139)	-.09(.044)		+.02(.058)	.79	2.28
6. OK*	-.45(.059)	+.65(.138)	-.06(.047)		+.00(.047)	.81	1.82
7. OK	-.52(.159)	+.57(.143)	-.08(.050)	+.10(.241)	+.17(.365)	.79	2.42
8. OK*	-.49(.129)	+.65(.142)	-.06(.053)	+.06(.203)	+.10(.306)	.81	1.78
1. MS	-.34(.031)	+.47(.069)	-.06(.026)			.90	1.51
2. MS*	-.34(.039)	+.47(.075)	-.04(.023)			.91	1.86
3. MS	-.36(.034)	+.47(.069)	-.05(.026)	+.037(.036)		.90	1.50
4. MS*	-.36(.044)	+.47(.070)	-.04(.023)	+.047(.047)		.91	1.86
5. MS	-.33(.033)	+.47(.067)	-.05(.026)		-.04(.027)	.91	1.54
6. MS*	-.32(.040)	+.47(.076)	-.04(.023)		-.04(.036)	.91	1.85
7. MS	-.28(.081)	+.47(.069)	-.05(.027)	-.06(.103)	-.08(.078)	.91	1.62
8. MS*	-.30(.110)	+.47(.079)	-.04(.025)	-.03(.138)	-.06(.104)	.91	1.86
1. LA	-.27(.028)	+.51(.060)	-.04(.029)			.89	1.69
2. LA*	-.27(.031)	+.51(.058)	-.04(.029)			.89	1.86
3. LA	-.31(.036)	+.52(.057)	-.05(.028)		+.07(.042)	.91	1.97
4. LA*	-.31(.036)	+.52(.059)	-.05(.030)		+.07(.042)	.91	1.90
5. LA	-.30(.028)	+.52(.055)	-.05(.027)	-.05(.026)		.91	2.03
6. LA*	-.30(.028)	+.52(.056)	-.05(.026)	-.05(.029)		.91	1.92
7. LA	-.27(.051)	+.52(.056)	-.05(.028)	-.10(.086)	-.08(.130)	.92	2.01
8. LA	-.27(.051)	+.52(.059)	-.05(.029)	-.10(.082)	-.08(.134)	.92	1.91

NOTE: A single asterisk means that an adjustment in the estimation for first order serial correlation was made using the maximum likelihood procedure in the Statistical Analysis System Proc Autoreg.

ESTIMATED ELASTICITIES FROM
VARIOUS STUDIES

Table A.II

Estimated Elasticities from Various Studies

Item

1	Blattenberger-Nation -.03
2	Lin-Northeast -.11
3	Beirlein-Northeast -.23 -.24 -.35
4	Herbert-Northeast -.22
5	Herbert-Northeast [-.52 -1.28] [- .56 -.92]
6	Herbert-Region -.30 -.39 [-.39 -.77]
7	Grady-Regions -.09* -.03* -.14 -.15 -.16 -.16
8	Herbert-This study [-.27 -.49]$_{LA}$ [-.36 -.59]$_{MS}$ [-.44 -.50]$_{NM}$ [-.46 -.62]$_{OK}$ [-.85 -.91]$_{AK}$

Sources: 1: Blattenberger, Taylor, and Rennhack, "Natural Gas Availability"; 2: Lin, Chen, and Chatov, "The Demand for Natural Gas"; 3: Beierlein, Dunn, and McConnon, "The Demand for Electricity"; 4: Herbert, "Data Matters"; 5: Herbert, "A Data Analysis and Bayesian Framework"; 6: Herbert, "Demand for Natural Gas"; Herbert, "A Data Analysis of Residential Demand"; and Herbert, "Measurement Error"; 7: Grady, "Regional Demand"; 8: Herbert, *Clean Cheap Heat*.

Notes to Table Items:
Item 1: The elasticity is obtained using an estimated marginal price of NG series. All other studies use the average price of NG. However, all indications are that an annual, state-level marginal price of NG series can be expected to be

highly correlated with an annual, state-level average price of natural gas series. The equation was estimated by a variance components technique with a constant state and temporal variance component for all states. The estimated equation included a one-period lagged dependent variable coefficient which was highly significant, with a magnitude of 0.92. Other estimated coefficients included an income (INC) elasticity, a marginal price of electricity (MEP) elasticity, a heating degree day (HDD) elasticity, and an indicator variable coefficient for 1974. The t-ratios were 1.66 and 1.93 for INC and MEP, respectively and thus INC might be considered as insignificantly different from zero. The HDD coefficient was highly significant as was the lagged dependent variable. The R^2 statistic was equal to 0.94. The data covered forty eight states for the years 1960-1974. The dependent variable was NG demand per customer.

Item 2: The equation was estimated by a general variance component technique in which the RNG demand equation was just one of several fuel/sector equations, which were estimated jointly. The equations were also weighted by the number of customers in the different states. The estimated average price of electricity was significant and positive at the 1 percent significance level. The INC elasticity was positive and was reported as insignificantly different from zero at the 1 percent significance level and at the 5 percent significance level. Other coefficients estimated were a price of oil (PO) elasticity, which was estimated to have the wrong or negative sign, yet to be significantly different from zero and a lagged dependent variable which was significant with a value of 0.96. The R^2 statistic was equal to 0.96. The dependent variable was NG demand per customer.

Item 3: The equation was similar to that used by Lin et al. cited in Item 2, except that the equations were not weighted by the numbers of customers in the different states, and six, rather than nine behavioral equations were estimated jointly. The dependent variable was NG demand per capita. The first, second, and third coefficients reported in the table were, for an ordinary least squares estimation, a variance component estimation in which sectoral equations were estimated separately and a variance component estimation in which the sectoral equations were estimated jointly. Other coefficients estimated were an EP elasticity, a PO elasticity, and a lagged dependent variable coefficient. The PO elasticity had the wrong sign. The INC elasticity ranged from 0.01 to 0.03 and was not significantly different from zero. The EP elasticity ranged from 0.17 to 0.32 and was significantly different from zero. The lagged dependent variable coefficient was equal to or greater than 0.9 in all instances and was highly significant. The R^2 statistic was equal to 0.99. The years examined were 1967-1977.

Item 4: After testing whether state and time period variance components were equal to zero, which hypothesis could not be rejected, a residential demand equation was estimated for the six major northeastern states, by ordinary least squares techniques, for the years 1970-1982. The INC and EP elasticities were equal to 0.10 and 0.17 and were significantly different from zero at the 1 percent significance level. The lagged dependent variable coefficient was only equal to 0.61, due to the inclusion of a proxy variable for effective space heating equipment stocks of the average RNG customer, which proxy variable captures differences in demand between states which, otherwise would have been captured by the lagged dependent variable. This proxy variable was highly significant. The other variable included was an indicator variable for 1973, 1974, and 1979. The dependent variable was NG demand per customer.

Item 5: A regression equation was estimated similar to Herbert, "Data Matters." A lagged dependent variable, however, was not included in this estimation. The implications of random errors-in-variables for the estimation was indicated by reporting maximum likelihood bounds, indicated by a bracketed number in the table, rather than point estimates for price elasticities. The first set of bracketed numbers in the table was obtained without making any assumptions about the proportion of the observed variance due to measurement error. The second set was obtained after making assumptions, based on knowledge of the data, about the proportion of the observed variance due to measurement error. The lower bounds for the PE and INC elasticities in the final estimation were equal to 0.21 and 0.15, respectively. The R^2 in the ordinary least squares estimation was equal to 0.92. The dependent variable was NG demand per customer.

Item 6: The first listed price elasticity reported was estimated in Herbert, "Demand for Natural Gas." A general variance components technique was used in which separate variance components were estimated for each state and covariance components were estimated for each pair of states. A correction was also made for first-order serial correlation. A lagged dependent variable was not used as an explanatory variable. Other explanatory variables included a proxy variable for effective space heating equipment stock, an indicator variable for 1974, and a variable to represent the cumulative conservation and other energy saving investments put in homes after 1978 that were reported for tax credit purposes. The EP elasticity was equal to 0.08 and was significantly different from zero at the 1 percent significance level. An INC variable was not included in the estimation. State data for the states of Missouri, Illinois, Indiana, Ohio, Michigan, West Virginia, and Kentucky were used for the years 1960-1982. The R^2 statistic was equal to 0.87. The second listed elasticity was estimated in Herbert, "A Data Analysis," and was otained using ordinary least squares techniques rather than variance component techniques because when fixed state effects were used, significant variance components and first-order serial correlation coefficients were not identified. The estimated equation was also evaluated in J. H. Herbert and P. S. Kott, "Robust Variance Estimation in Linear Regression," *Journal of Applied Statistics*, 15 (1988): 341-345. The data set was similar to that used in Herbert, "Demand for Natural Gas." Data for Illinois was not included in these estimations. The R^2 statistic was equal to 0.96. The EP elasticity was equal to 0.15 with a t-ratio equal to 12. The final bracketed set of estimates reported is the range of maximum likelihood estimates for the true regression coefficient obtained without making any specific assumptions about the degree of random measurement error in any of the variables reported in Herbert, "Measurement Error." The dependent variable in all instances was equal to NG demand per customer.

Item 7: Grady estimated NG demand equation by variance component techniques, conducted tests of functional form, and weighted the estimated regression equations by the number of NG customers in the different states. Grady grouped the states into regions based on the ratio of the total miles of NG pipelines to the state area, total miles of pipeline per household, and the amount of NG produced in the state per household. The data included the years 1968-1978. The smallest number of states in a group was five. As indicated by an asterisk, several of the price elasticities were not significantly different from zero. Coefficients were also estimated for an EP variable which was found to be negative for two regions and to be positive and significantly different from zero at the 5 percent significance level for one region; and an INC variable which was found to be negative for five of six regions. Other variables included were HDDs, a lagged dependent variable, and the ratio of NG households to total households. The dependent variable was NG demand per customer.

Item 8: The bracketed numbers are bounds for the coefficients of the state equations reported in chapter 8 of this book.

A NEW ECONOMIC MODEL
BASED ON BALESTRA'S MODEL

The primary relationships in the Balestra and Nerlove (B&N) model are
First,

$$G_t - (1-r_g)G_{t-1} = B_0 - B_1p_t + B_2[F_t - (1-r)F_{t-1}] \quad (1)$$

$$G_t = B_0 - B_1p_t + B_2[F_t - (1-r)F_{t-1}] + (1-r_g)G_{t-1} \quad (2)$$

$$G_t = B_0 - B_1p_t + B_2[F_t - (1-r)F_{t-1}] + B_6G_{t-1} \quad (2a)$$

where

G	= demand for NG normalized by weather,
r_g	= depreciation rate for NG appliances,
p	= price of NG normalized by consumer price index,
F	= aggregate demand for all fuels normalized by weather (either a surrogate for space-heating requirements or for the stock of all appliances),
r	= constant depreciation rate for non-gas appliances,
t	= time index.

Second,

$$F_t - (1-r)F_{t-1} = uW_t - (1-r)uW_{t-1} \quad (3)$$

where

W	= stock of appliances,
u	= a constant rate of utilization because of a high "and, therefore, constant" efficiency of combustion.

Third,

$$F_t = B_3 + B_4 N_t + B_5 Y_t \tag{4}$$

where

N = population,
Y = income.

Equation (4) is viewed as an empirical regularity which has been established by many prior estimations of regression equations. The B&N formulation could be stated alternatively either in appliance demand or in fuel demand terms because of the assumption of a constant rate of utilization. When equation (4) is substituted into equation (2), it yields the model,

$$G_t = B_0 + rB_2 B_3 + B_1 p_t + B_2 B_4 N_t + B_2 B_5 Y_t \\ + rB_2 B_4 N_{t-1} + rB_2 B_5 Y_{t-1} + B_6 G_{t-1}. \tag{5}$$

Equation (5) indicates that differences in energy demand between time periods are a consequence of changes in the number of energy-using appliances between time periods or a simple stock adjustment process. NG demand is a function of the capability of the NG industry to capture its replacement market as determined by the appropriate depreciation rates, and the new market for energy appliances as determined by population growth and income growth. In the B&N Model, weather is controlled for by normalization. Income does not affect usage. B_2 represents the proportion of the new residential energy market captured by NG. B_5 represents the proportionate relationship between income level and number of appliances (or fuel demand). Price determines relative market shares but not relative usage. Capital equipment efficiency is viewed as constant and independent of price.

THE CURRENT ECONOMIC MODEL

Each residential customer is assumed to own only one space heating appliance (unit).

Let

GS_{t-1} = gas space-heating customers in year t-1,
$r_g GS_{t-1}$ = gas space-heating customers who must replace a furnace between year t-1 and year t,
$a(r_g GS_{t-1})$ = gas space-heating customers who replace an old gas furnace with a new gas furnace between year t-1 and year t,
FS_{t-1} = non-gas space-heating customers in year t-1,
$(1-r_g)GS_{t-1}$ = NG space-heating customers from year t-1 remaining in year t,

$b(FS_t-(1-r)FS_{t-1}) =$ new market for space-heating customers captured by NG industry between year t and year t-1,

Thus, the number of NG space-heating customers in year t is defined as:

$$GS_t = a(r_g GS_{t-1}) + b(FS_t - (1-r)FS_{t-1}) + (1-r_g)GS_{t-1}. \qquad (6)$$

This definition for number of NG space-heating customers (appliances) is an analogue for the B&N general model, except that the proportions a and b are represented explicitly, rather than being a function of the price of NG and other factors in some year t.

In the B&N formulation, gas demand (GAS) was normalized by heating degree days (HDD). In the present formulation, HDD and GS are moved to the right-hand side of the equals sign, and both GAS and GS are normalized by the number of customers (CUS):

$$GAS/CUS = f_x([GS/CUS] \cdot HDD) = f_x(HR), \qquad (7)$$

where f_x is a function that depends on other factors in addition to HR. The variable HR denotes space heating requirements. There are distinct advantages in using the HR variable rather than just heating degree days (HDD) as is usually done. Space-heating customers consume more NG than non-space-heating customers. In a period of rising NG prices and of a rising proportion of gas space-heating customers, for example, there will be a tendency to understate the effect of price if changes in the proportion of space-heating customers are not controlled. The HR variable also captures more of the between-state variability in gas sales than would otherwise be the case.

Appendix IV

DATA USED TO ESTIMATE EQUATIONS
IN CHAPTERS 2 AND 3

State		1910	1911	1912	1913	1914	1915	1916	1917	1918	1919
WV	QD	157.15	158.62	171.64	153.36	157.89	154.37	151.58	164.41	164.89	148.09
	PG	17.40	18.12	17.49	18.21	17.34	16.89	15.46	13.72	13.05	12.32
	DT	1.02	1.04	0.94	1.06	0.98	0.98	1.00	0.94	1.06	1.02
OK	QD	139.13	129.69	138.25	142.76	123.50	148.15	134.50	136.07	178.23	139.43
	PG	16.90	16.90	19.11	16.83	17.80	16.55	15.30	14.79	12.65	14.97
	DT	0.98	1.00	0.97	0.89	1.14	0.86	1.16	0.95	1.02	0.96
PA	QD	135.04	124.12	142.67	116.51	121.61	116.52	120.04	131.40	124.33	117.58
	PG	25.30	26.02	23.80	25.29	25.40	25.13	23.34	20.92	19.48	18.22
	DT	0.98	1.04	0.94	1.06	0.96	1.00	1.00	0.94	1.09	1.02
OH	QD	127.32	100.11	104.64	94.37	98.02	99.66	101.16	116.49	110.65	101.71
	PG	25.00	27.40	27.92	27.17	27.00	27.03	25.50	22.58	20.09	18.43
	DT	0.98	1.04	0.94	1.06	0.96	1.00	1.00	0.94	1.09	1.02
NY	QD	114.96	115.89	117.99	110.00	116.13	112.42	117.74	126.21	115.98	99.23
	PG	29.70	30.39	28.87	28.67	28.71	28.73	26.73	22.91	20.86	19.84
	DT	1.00	1.00	0.98	1.06	0.95	1.02	0.98	0.96	1.06	1.02

State		1920	1921	1922	1923	1924	1925	1926	1927	1928	1929
WV	QD	161.34	140.06	141.65	148.03	155.69	143.56	155.01	136.58	147.98	134.98
	PG	25.30	31.68	35.26	35.46	34.69	36.91	36.00	37.85	38.60	40.47
	DT	0.98	1.08	0.96	0.98	0.94	1.06	0.96	1.04	0.98	0.98
OK	QD	159.74	130.42	127.40	122.61	114.43	106.51	105.71	115.14	107.34	138.91
	PG	30.50	40.52	51.87	56.59	62.11	60.69	56.83	54.46	59.53	52.16
	DT	1.04	1.05	0.98	0.97	0.98	1.04	0.98	1.02	0.96	0.91
PA	QD	129.68	114.35	102.95	107.52	103.65	95.78	93.60	84.81	86.08	83.92
	PG	37.30	48.81	58.80	58.94	63.16	68.57	68.37	69.92	72.05	71.93
	DT	0.96	1.08	0.96	0.98	0.96	1.04	0.96	1.04	0.96	1.06
OH	QD	103.21	87.31	85.52	89.58	82.81	76.53	77.99	72.76	74.22	72.62
	PG	38.30	47.46	59.64	59.65	64.10	64.00	69.74	71.77	73.33	74.62
	DT	0.96	1.08	0.96	0.98	0.98	0.96	1.04	0.96	1.06	0.96
NY '	QD	93.72	87.38	86.83	81.04	74.67	72.08	71.84	63.47	67.83	67.83
	PG	39.00	48.02	58.69	68.22	69.49	70.29	73.70	77.42	80.12	78.71
	DT	0.96	1.06	0.98	0.98	0.98	1.02	0.96	1.04	1.02	1.00

Sources: Quantity and Price, *Mineral Resources*, 1910-1924, and *Mineral Yearbook*, 1925-1929, various pages; Deflator, Consumer Price Index All Items, *Historical Statistics, Census*, 211; Temperature data, *Historical Statistics, Census*, Series J, 441-447.

Note: QD is average consumption per customer in 1000 cubic feet; PG is average price per 1000 cubic feet in constant dollars where values from 1910 to 1919 are indexed to 1910 dollars and values from 1920 to 1929 are indexed to 1920 dollars; DT is the average temperature for a year divided by the average temperature for the previous year. Numbers are rounded to nearest one-hundreth.

BIBLIOGRAPHIC ESSAY

To list all the sources already cited in the text would be a duplication. However, there are a few sources that require special attention and a few that were not cited but were influential in determining the direction of the book.

I initially learned much about the importance of data analysis for analyzing the development of markets from reading: Schultz, H., *The Theory and Measurement of Demand*. Chicago: University of Chicago Press, 1938.

Publications prepared under the auspices of the National Bureau of Economic Research demonstrated how the reporting of industry detail and the full documentation of primary data could greatly aid additional research. Barger, H. and S. H. Schurr, *The Mining Industries, 1889-1939*. New York: National Bureau of Economic Research, 1944. Gould, J. M., *Output and Productivity in the Electric and Gas Utilities, 1899-1942*. Cambridge: University Press, 1946.

For an understanding of the significance of the natural gas industry prior to 1935, there is no better source than the 96 volume FTC study of the public utility corporations. Federal Trade Commission. *Report to the Senate on Public Utility Corporations*. Senate Document No. 92, 70th Congress, 1st Session.

Other, more recent and much briefer studies were also found to be useful for a history of the NG industry up to the 1970's and for an understanding of perceptions about the expected direction of the industry for the 1980's: Federal Power Commission, *Natural Gas Survey, Vol. 1, The Commission Report*. Washington, D.C.: U.S. Government Printing Office, 1978. Federal Power Commission, *Natural Gas Survey, Vol IV, Distribution*. Washington, D.C.: U.S. Government Printing Office, 1973. Federal Power Commission, *Natural Gas Survey, Vol V, Special Reports*. Washington, D.C.: U.S. Government Printing Office, 1973.

There were several sources that proved especially fruitful for understanding the intensity of the political conflicts and the interaction of government, citizens, and business in determining the terms for the regulation of the gas and related industries. Sanders M. E., *The Regulation of Natural Gas, 1938-1978*. Philadelphia, Pa.: Temple University Press, 1981. Clark, J. G., *Energy and the Federal Government, Fossil Fuel Policies, 1900-1946*. Chicago: University of Illinois Press, 1987. Goodwin, C. D., ed. *Energy Policy in Perspective: Today's Problems, Yesterday's Solution*. Washington, D.C: The Brookings Institution, 1981.

For a rich analysis of social processes and technological change one work stands out. Rose, M. H., "Urban Environments and Technological Innovation: Energy Choices in Denver and Kansas City, 1900-1940." *Technology and Culture* 25 (1984): 503-539.

INDEX

About the Author

JOHN H. HERBERT is Senior Economist with the Energy Information Administration and an adjunct professor of statistics at Virginia Polytechnic Institute and State University. He has spent more than fifteen years as an energy consultant in both the public and private sectors, with ten years devoted to the natural gas industry specifically.